Praise for *Strange Survivors*

"*Strange Survivors* presents some of the most remarkable examples of animals' survival mechanisms in an accessible, scientifically accurate, and humorous way. Reading this book was like sitting down to enjoy a chat on the wonders of the natural world with a close friend."
—MARIE MCNEELY, PHD, cofounder of
People Behind the Science and Unfold Productions

"*Strange Survivors* is technically accurate and isn't shy with scientific terms, but never in a manner that's intimidating or overbearing. This isn't fiction—you have to think as you read, but the reward is an increased understanding of the astonishing variety of life on Earth and the strategies species use to survive."
—PETER CAWDRON, author of *Anomaly*

STRANGE
SURVIVORS

STRANGE SURVIVORS

HOW ORGANISMS ATTACK AND DEFEND IN THE GAME OF LIFE

ONÉ R. PAGÁN

BenBella Books, Inc.
Dallas, TX

About the cover image: Mantis shrimp use natural ballistics to capture their prey with ultrafast spiny appendages—they also have unique visual systems that allow them to see as many as 100 million colors (humans can see between 1 and 10 million). Image courtesy of worldartsme.com.

BenBella

BenBella Books, Inc.
10440 N. Central Expressway, Suite 800
Dallas, TX 75231
www.benbellabooks.com
Send feedback to feedback@benbellabooks.com

Printed in the United States of America
10 9 8 7 6 5 4 3 2 1

Library of Congress Cataloging-in-Publication Data:
Names: Pagán, Oné R., author.
Title: Strange survivors : how organisms attack and defend in the game of
    life / One R. Pagan.
Description: Dallas, TX : BenBella Books, Inc., [2018] | Includes
    bibliographical references and index.
Identifiers: LCCN 2017052642 (print) | LCCN 2017058961 (ebook) | ISBN
    9781944648596 (electronic) | ISBN 9781944648589 (trade paper : alk. paper)
Subjects: LCSH: Drug resistance in microorganisms. | Antibiotics--Development.
Classification: LCC QR177 (ebook) | LCC QR177 .P34 2018 (print) | DDC
    616.9/041--dc23
LC record available at https://lccn.loc.gov/2017052642

Editing by Laurel Leight and Alexa Stevenson
Copyediting by Scott Calamar
Proofreading by Sarah Vostok and Amy Zarkos
Indexing by Jigsaw Information
Text design and composition by Silver Feather Design
Front cover design by Faceout Studio
Jacket design by Sarah Avinger
Printed by Lake Book Manufacturing

Distributed to the trade by Two Rivers Distribution, an Ingram brand
www.tworiversdistribution.com

Special discounts for bulk sales (minimum of 25 copies) are available.
Please contact Aida Herrera at aida@benbellabooks.com.

To Liza.
Twenty-five years and counting!

CONTENTS

Introduction 1

CHAPTER 1
The *E* Word 7

CHAPTER 2
The Language of Life 23

CHAPTER 3
It All Starts with a Spark 43

CHAPTER 4
Unusual Suspects 71

CHAPTER 5
The Fast and the Hangry 113

CHAPTER 6
The Very Best Survival Tactic of Them All 145

Postscript 173
Acknowledgments 175
Notes 177
Bibliography and Further Reading 193
Index 217
About the Author 229

INTRODUCTION

I want the very first words you read from me to be an expression of how grateful I am that you chose this book. This is a wonderful time to be a science enthusiast; there are a plethora of popular science books for you to choose from—and yet, here you are, reading mine! I hope you will enjoy it, and I thank you from the bottom of my heart.

Why should you read this book? In a nutshell, because the living world is majestic and awe inspiring, and because by learning about it we may also come to understand ourselves a bit better. It is as simple as that. Life on Earth is—still, despite the breakneck pace of extinctions—majestically diverse, and displays, in Charles Darwin's most famous words, *"endless forms, most beautiful and most wonderful."* Biological life fills us with wonder with its unending mysteries, puzzling behaviors, breathtaking beauty, raw violence, and unexpected liaisons. However, beauty, majesty, and wonder are only part of the proverbial picture. Virtually every peculiar trait, every behavior, every colorful characteristic that makes the biological world so interesting to look at and entertaining to explore evolved to confer practical benefits to the organisms displaying it, and many of those traits also offer us tools to improve our own chances of survival—whether directly, as with substances that prove useful in curing disease or otherwise improving human life, or indirectly, as knowledge that informs technological, engineering, or medical research. You see, we do not have to choose between wonder and practicality; with nature, we can certainly have both.

We humans are counted among those endless, beautiful forms Darwin spoke of. Moreover, whether we realize it or not, whether we like it or not, we are intimately connected to all the other life forms with which we share this planet. There is one characteristic in particular that we share with all life, a drive so fundamental that it is at the root of much of what we do and are. This drive has had many names over time, such as the will to live or preservation instinct—personally, I prefer to call it the *survival instinct*. This instinct is an undeniable biological imperative, and because of it, life has evolved a mind-boggling variety of strategies to stay, well, *alive*—to survive long enough to transmit its genetic information to the next generation.

I will say this in various ways throughout the book, but the main point is that all organisms have a primary "mission," which can be divided into three parts: (1) to get energy from the environment, (2) to avoid becoming food for other organisms, and (3) to pass on their genetic information. From the perspective of the biological world, these are the most—nay, the *only*—things that matter.

Before we get started, a few details about this book: First, I have endeavored to be as accurate as possible, meaning that whenever I was able, I've drawn stories directly from the original scientific literature or from reliable books on the topic. I've also made liberal use of both footnotes* and endnotes to offer additional resources, more information on specific aspects of the chapter in focus, and amusing (if only tangentially relevant) anecdotes for your reading pleasure. Speaking of tangents, when an interesting detour seemed both too distracting to allow yet too fascinating to deprive you of, I have set it off in a box, so that you can delight in the foliage without losing sight of the forest. At the end of the book, I've included a bibliography with both general sources as well as more specialized references should you wish to delve further into any of the subjects we cover (the bibliography also includes a few goodies for your reading

---

* Like this one.

pleasure). The natural world is so rich that there is simply no space in a normal-sized book to do it justice. One group of organisms that I had to leave out, except in the most fleeting way, is plants—not because they are uninteresting or lack diversity (if anything, the opposite is true), and certainly not because they are passive bystanders in the battle for survival. Plants are master survivors, and far from passive: they can warn each other of danger, recognize their relations, and even have a memory of sorts. By all means, do take the time to learn more about them!*

A note about Chapter 2, "The Language of Life," in which I explore questions of what life is and what makes it tick: These concepts play a central role in subsequent chapters and will help you to better understand the forces underpinning the survival strategies we talk about. That said, if you are conversant in these matters, please feel free to skip ahead and enjoy the rest of the book. And while in any given chapter I may refer to information presented in others to emphasize a point or to allude to curious connections, they are otherwise quite independent from each other, and you need not necessarily read them in the order they appear. This is, after all, your book. Read it any way you like. You do not even have to read the rest of this introduction, if you don't want to!

But, just in case you *do* want to, I will go ahead and finish it up.

I wrote this book with the semi-mythical "interested layperson" in mind. This means that I will not be excessively technical, but neither will I be patronizing. This book is for you, my fellow science enthusiast. I was a bookworm long before I became a science professional, and regardless of our respective walks of life, we belong to the same club. I've tried to write as if we were having a conversation over coffee (assuming your coffee conversation focuses on things like venoms and killer snails). Ideally, this conversation will be interesting,

---

* A recent book that offers fascinating insight into the life of plants, along with insight into the life of a scientist, is Hope Jahren's excellent memoir *Lab Girl*.

thought-provoking, and—why not?—amusing as well. The natural world is certainly awe-inspiring, but it is also deeply weird and often frankly hilarious: this is part of its considerable charm.

I would be remiss if I did not acknowledge the influence that blogging has had on my writing in general and this book in particular. Blogging is how I tasted for the first time the elation of presenting my informal musings to an audience. My blog, *Baldscientist*, was the initial repository of a few of the stories that you'll see in *Strange Survivors*, though of course these have been further developed and significantly expanded for this book.

Finally, virtually every popular science writer aspires, whether consciously or not, to reflect the style of an admired writer or scholar, and I am no exception. I admire a great many science writers and their books, but I wish to acknowledge two books in particular that inspired the writing of my own: *Mother Nature Is Trying to Kill You* by Dr. Daniel Riskin, and *Animal Weapons: The Evolution of Battle* by Dr. Douglas J. Emlen. Each takes a different approach to exploring how organisms evolved their defensive and offensive strategies, and each informed my thinking as I shaped my own approach for this book.

There is another author I wish to acknowledge, one who has inspired in countless souls a profound awe of biological life in all its endless and most wonderful forms. This author is nature herself. Every single day, nature writes, revises, erases, and re-creates aspects of life, and much of this work we will never even know about— either because it is invisible or simply goes unnoticed—much less understand. What nature creates is not perfect, because after all nothing is, but perhaps nature's greatest and most unexpected skill is to transform literal imperfections—an accident during cell division, a mutation of the genetic material, among many other instances— into life-changing advantages. Nature delights us with the most unexpected plot twists, surprising and intriguing, twists no human author could ever imagine. I'd like to think that nature loves mysteries, and it is a wondrous fact that these mysteries are in principle explainable by applying the methods of science and mathematics. In my mind, the most interesting of these mysteries are (1) how matter came to be, (2) what time "is," exactly, (3) how matter organized itself from "mere" physics and chemistry into what we call life, and (4) how life organized itself to allow consciousness to emerge, again from the building blocks of physics and chemistry. The first two mysteries are above the pay grade of most scientists I know (they are certainly above my pay grade). As for the last two, well, I'd say we are making good progress.

As far as we know, we seem to be the only creatures on this planet with the type of consciousness that is curious enough to ponder mysteries like the ones above. This being said, I wholeheartedly agree with Darwin's* contention that the difference between human and animal intelligence is one of "degree rather than kind." We'll probably never know whether a chimpanzee, standing by the shore of a peaceful lake, watching the sunset, wonders what the brilliant circle of light slowly descending over the horizon might be, and we have even less hope of knowing what a fish in the same lake thinks while noticing the looming darkness. I, for one, am grateful for the opportunity to contemplate and even understand some of nature's mysteries. We'll talk about a few of these mysteries in this book, namely those surrounding some of the strategies that life uses to play and win in a game that began billions of years ago, the world's original game, and the one with the highest stakes, for it is literally life or death: the game of survival. Will you join me?

—ONÉ R. PAGÁN
October 2017
Somewhere in PA

---

* I am an unapologetic "Darwinophile."

CHAPTER 1

THE *E* WORD

I am sure that you have already guessed the word hinted at in the title of this chapter. If you immediately thought "evolution" you are, of course, correct! The concept of organic evolution is the unifying theme of all biology; there are no exceptions to this rule. Thus, not surprisingly, evolutionary processes play a central role in how and why living organisms display offensive or defensive strategies. If we were to compare biological evolution with a boat upon the water, life would be the vessel's engine, and death would be its rudder. Virtually every type of evolutionary strategy that we know of implies the avoidance of the death of the individual (or of closely related individuals, particularly in large societies of organisms), at least for as long as possible—with the inherent goal of living as long as necessary to achieve reproduction. Alas, death is inevitable, although it seems that we humans are the only beings aware of this eventuality.

The certainty of death* and the instinct to avoid it drive the selection for survival of specific organisms in a population that, over time, may change the species as a whole. It is important to note that properly stated and understood, the theory of evolution does not pretend

---

* Even though we seem to be the only type of organism on this planet that is aware of our mortality, all organisms fight to survive.

to explain the origin of life, not even implicitly. Biological evolution only began once proper life was present on Earth. Despite—and probably because of—its central importance to a true understanding of the nature and diversity of life on this planet, "evolution" is the most misused, abused, misapplied, and grossly misunderstood word in all of biology, bar none. There is no other biological term capable of eliciting such passionate reactions from so many diverse people.

Even though evolutionary theory is traditionally considered a historical science, the fundamental concepts related to the process of biological evolution are no mere academic curiosities; they have real-world applications. Evolution is essential to understanding life's history on our planet; there's no question about that. Furthermore, a working knowledge of evolution and its implications is of particular importance in understanding a few literal "clear and present dangers" to the welfare of our society at large. Just to give you an example, consider the emerging problem of antibiotic resistance. In brief, many infections that, sixty or so years ago, were easily treatable with antibiotics, are no longer so easily treatable because the bacteria that cause these infections have developed resistance against such medications. Individual bacteria possessing a trait that, for whatever reason, inured them to the medication's effects were in turn the bacteria that survived and reproduced. As a result, over time, bacteria with these traits became more and more common, while the more readily killed bacteria were, well, readily killed. In essence, the medicines do not work anymore because wily bacteria have evolved to survive them. This is an increasingly alarming public health problem with no clear solution in sight. Evolutionary science may help us understand and eventually solve it.

However menacingly death looms in the distance, life is ever resourceful. As we study evolution, and as we keep unraveling its mysteries, we identify the numerous schemes that organisms use to participate in the unending game of biological life. In this book we will explore some of the most fascinating, and, frankly, weird examples of evolution in action—from electric catfish to regenerating worms to toxic birds to spitting spiders! One cannot help but marvel at the mechanisms through which evolution came up with survival strategies such as these.

So what exactly is "biological evolution"?* In a nutshell, it is biological change over time. It is that simple. It is clear that there are life forms existing today that did not exist, say, ten million years ago. We are intimately acquainted with one such life form: ourselves! We, as a species, have been around for a mere two hundred thousand years or so, depending on whom you ask and on how you define what a "human" is. Before that time, our ancestors were living beings quite similar to us, but not exactly like us; in fact, the farther we go back in time, the more different our forebears are from us. Humans are an evolved species, and we are still evolving. This is true of every living species on Earth right now; we are only one example of the majestic biodiversity of life. In a literal sense, each form of life living on this planet is a testament to the exquisite adaptations that are indispensable for survival.

In his 1995 book *River Out of Eden*, science popularizer Richard Dawkins famously stated an obvious yet very profound truth: not one of your ancestors—or mine, for that matter (in fact, many of these may be one and the same)—died childless. This is true of each and every organism that you have ever seen or that you may see in the future. We are all part of an unbroken chain of beings who survived long enough to reproduce: true winners in the game of life. In biology you are judged a winner not by your beauty or your strength or your smarts, but by the number of your surviving offspring. That's the only way of keeping score that actually matters.

Yet for every single example of successful life, there are millions of examples of lineages that no longer exist today. In the best-case scenario, one reason for the disappearance of a lineage is that it evolved into another. However, the most common occurrence is that

---

* And from now on, unless I tell you otherwise, whenever I say "evolution," I will mean "biological evolution." Agreed?

biological lineages disappear forever when organisms die without descendants. One of the best illustrations of this fact is the ample evidence documenting the existence of life forms that were around in long bygone eras, yet are not around anymore. Extinction is a fact of life. It has happened before, and it keeps happening; no serious scientist denies this.

Probably the best-known type of evidence for this fact is the wide variety of fossils discovered over the years by dedicated paleontologists as well as enthusiastic amateurs. Many of these fossils tell us parts of the stories of organisms that are no more. When we think "fossil," the most likely image that comes to mind is, of course, those terrible lizards, the peerless dinosaurs. There is, for example, everyone's favorite, *Tyrannosaurus rex*, the undisputed "king" of the dinosaurs; there's also the *Brontosaurus*, whose very existence was for a while cast in doubt[1]—but at least the name is coming back; the *Stegosaurus*; and the *Triceratops*, among many others. There's no question that more than one scientific career began with a love for dinosaurs.

WHAT'S IN A NAME?

Whenever describing or even mentioning an organism in this book, in addition to its common name if available, I will also give you its scientific name, which includes its *genus* and its *species*. This is the method that biologists use to systematically classify life (there are other categories, but we need not concern ourselves with them here). For example, *Tyrannosaurus rex* is the scientific name for the most famous dinosaur ever, while the scientific name for humans is *Homo sapiens*. As shown, the whole name is italicized, the genus is capitalized, and the species is all in lowercase letters. The advantage of using scientific names is that sometimes there can be more than one common name for a particular organism, which can be confusing. Case in point: the common names "Gallipato," "Ofegabous," "Ribbensalamander," and several other names refer to exactly the same animal, *Pleurodeles waltl*, which is a certain type of newt. As you can see, the use of scientific names avoids confusion and quite simplifies the identification of organisms.

We know of the terrible lizards mostly by their skeletons, but we also have other remnants of them, namely skin impressions, and we have some of their tracks, too, as well as outlines of the whole dinosaur. This helps to illustrate the fact that as far as fossils are concerned, we are not limited to examining petrified bone. Furthermore, for smaller organisms like insects, small frogs, and lizards, we sometimes have whole bodies conserved in amber. Speaking of which, in December 2016, an interdisciplinary team reported the finding of a piece of a small dinosaur's tail encased in amber, and it had feathers![2] The adventure of discovery never ends.

Sadly—at least from the perspective of anyone interested in nature in general and the history of life in particular—as rich as the fossil record is, it represents a mere fraction of the true biodiversity of Earth over time. Most estimates seem to indicate that less than one percent of all the life that has ever existed here has left traces in the fossil record. This is due to a variety of factors. One of them is that the process of fossilization requires a set of rather special circumstances. You see, scavengers such as insects, as well as decomposers like bacteria and fungi, among others, are literally everywhere and go immediately to work once an organism dies, harvesting the decaying tissue for their various purposes. Just by blind luck, some organisms are quickly covered by sediment soon after death. This may create the conditions necessary to preserve hard parts, like bones. These are the easy ones, so to speak. But then there are the many organisms that lack hard parts and are therefore harder to fossilize. Despite this, we do have quite a few examples of soft-bodied organisms in the fossil record. Unfortunately, there are animals that by pure bad luck (for us) of their physiology are next to impossible to fossilize. Case in point: my favorite organisms, the planarian worms. Most species of these worms die rather histrionically, as their bodies dissolve shortly after life escapes them. This happens to be the organism that I use in my own scientific research, and I have observed this phenomenon "in person" as it were. I like these worms very much. How I wish for a true example of a fossilized planarian!

In addition to the preserved remains of long-dead organisms, we also have many other lines of evidence of past life. Some of these lines of evidence include a close examination of the taxonomic relationships between organisms, as encrypted in the genetic code. You do not have to dig too deeply to find that within every single type of organism examined to date, there are indeed molecular fossils consisting of non-active DNA, ancient genetic material that tells us the stories of former physiologies and biochemistries, as well as stories of past battles with infectious organisms.

Ever since humans realized that life on this planet changes, we have concocted a variety of explanations that tried to account for its diversity. This was by necessity a "hit-and-miss" progression because as science advanced, we refined the data that we use to come up with such explanations. Depending on the source, the explanations offered to explain the diversity of life on our planet have ranged from the pretty good to the pretty silly, and in most cases such explanations closely reflect the sociological environment of the time in which they appear. Since 1859, the best available evidence has pointed to *natural selection*, as first outlined by a fellow named Charles Darwin,* as the most likely mechanism accounting for evolutionary change.

In the twentieth century, natural selection was reinterpreted in light of the new science of genetics, and mutations were considered the primary source of variation that led to selection in the great game of evolution. These mutations, alongside population sizes and

---

* This guy is everywhere in biology, and this is no understatement!

environmental influences, set the stage for the evolutionary process. However, as science further progressed and our understanding of the fundamental characteristics of life was refined, we found that there are other ways in which variation, and therefore evolution, can occur. Some of these are rather well-known. Phenomena like *symbiosis* (essentially cooperation between organisms of different species) and *genetic drift* (a change in the nature of a population brought about by random occurrences like natural disasters, among other accidents) are well-documented examples of sources of evolutionary change. In the last fifty years or so, additional research showcases other mechanisms that are thought to influence evolution at the molecular level. These include various molecular rearrangements of genetic material (such as "jumping genes," genome duplications, chromosomal translocations, and so on), horizontal gene transfer (the transfer of genetic material not necessarily related to reproduction), and epigenetics (when genes are selectively activated or inactivated in response to environmental factors). Some scholars have even theorized that the fusion of whole genomes from different species is a mechanism for evolutionary change! It would not be too surprising if other mechanisms as yet unknown also play a role in the process of evolution.

It is important to point out that multiple explanations for the fact of evolutionary change can coexist. In fact, there are well-documented cases where two or more of these mechanisms work together. Life does not always choose to do things one way and exclude another way. Please remember that we classify these mechanisms for our benefit, so we can make sense of life. Nature will work with all of them simultaneously. Life is that wonderful.

EVOLUTIONARY ARMS RACES

Many of the strange survival strategies that we will look at in this book can be explained by the concept of *evolutionary arms races* (sometimes called "biological arms races"). In its original conception, the idea of an "arms race" was political in nature. The process describes two countries competing to possess the best armies, the best equipment, and the best resources. Probably the best-known example is the Cold War, which was a state of rather tense "peace" involving the United States and the former Soviet Union, as well as their respective allies at the time. This period spanned roughly between 1945 and 1991. Truth be told, this "cold" war was somewhat "warmer" in some places, as the two superpowers waged war by proxy, as it were.[3] Briefly stated, arms races entail an increase in the manufacturing and accumulation of weapons and associated technology between the two factions. Whenever an improvement in the status of one nation occurs, it immediately triggers a swift response from the other nation. Over time, both sides will have more weapons and resources than necessary to defeat the perceived enemy. The expression that comes immediately to my mind is "overkill."

Many examples of biological arms races have been well documented, as this is one of the main engines of evolutionary change. This process seems to be an important source of selection pressure that triggers the biological change that we know as evolution. Probably some of the best-known examples of biological arms races involve the *coevolution* of predators and prey.

THE TALE OF THE CHEETAH AND THE GAZELLE

The cheetah is one of the several magnificent big cats that are—sadly—disappearing from our world. Zoologists think that there are five species of cheetahs, four of them from various regions of Africa and an additional species found in the Middle East and parts of Asia. Cheetahs are widely recognized as the fastest land animals, at least for short distances. The speed of the cheetah is generally explained as the product of a biological arms race between these predators and their main prey, which includes gazelles and antelopes, the cheetahs' peers as far as land speed is concerned.

In a given prey population, there will be individuals that are a tad faster than others in their herd. By virtue of their speed, these individuals will be more likely to escape predation and therefore more apt to survive and reproduce, which passes the genes responsible for their speed to the next generation. On the predatory side of the fence, slightly faster cheetahs will likely be more successful hunters, thus surviving and passing their "speed genes" to the next generation. The process goes through several iterations, generating faster cheetahs and faster gazelles over time.

Speed is not the only characteristic that may coevolve in the shared story of gazelles and cheetahs. Perhaps other characteristics like visual acuity, intelligence, and agility, as well as a variety of other factors, also contribute to optimize the hunting abilities of cheetahs and the escaping abilities of gazelles.[4] It is a safe bet that every single survival trait of these and other animals developed as a consequence of some kind of evolutionary arms race.

A RATHER WICKED MOLECULE

An interesting aspect of evolutionary arms races is that they are not limited to the "macro" world. Evolutionary arms races are quite common in the microbial and even the molecular world. There are examples of arms races between bacteria and viruses as well as intramural arms races between bacteria and bacteria, among other combinations. Moreover, there are undeniable arms races between parasites and humans, viruses and humans, and between bacteria and humans.[5] There are many examples that illustrate this process, including many that have been described at the molecular level. A particularly rich source of insight into this biological process arises from the study of how biological toxins guide evolutionary changes. This is one of the most fascinating topics in all of biology. In the words of a person of whom I am particularly fond:

"Nature is the best chemist. During the course of evolution, through literally millions of years, a wide variety of organisms have developed substances used for defense against predators, or to become predators themselves. As part of the evolutionary process, chemical structures beneficial for the survival of the organism are conserved; many of these molecules include small organic toxins."[6]

I must now confess to quoting myself. The above is the very first paragraph of my PhD dissertation. I hope that you do not perceive my citing myself as narcissistic. I include these words simply to express how fervent my interest is in this topic. I love toxins and venoms! I promise to talk more about toxins and venoms in general in an upcoming chapter, but in the meantime, let me tell you about a fiendish molecule called tetrodotoxin (TTX) and its central role in a well-documented example of a biological arms race.

Tetrodotoxin is a small molecule, but nonetheless quite lethal, with no known antidote against it. TTX was originally discovered in the 1950s, in the eggs of certain puffer fish species. However, people have known of the toxicity of puffer fish for quite a long time. There is evidence that multiple ancient cultures, including the Chinese and the Egyptians, were aware of the danger posed by this type of fish. In Japan, puffer fish is known as *fugu*, a traditional delicacy that dates from more than 1,500 years ago.* The toxin usually concentrates in the liver and in the eggs, which are removed from the fish during proper preparation. The dish must be prepared from the freshest fish available, because if too much time passes between catch and consumption, the toxin can find its way into the fish's meat. Over the years, several deaths have been attributed to improperly prepared fugu. Fugu enthusiasts seem to like the tingling sensation that they feel on their lips when eating it, a sensation directly related to the dish's tetrodotoxin content. In brief, TTX decreases the activity of

---

* Although strictly speaking, fugu comes generally from puffer fish, this dish can be prepared from about twenty different types of fish.

nerves around the lips, mouth, and tongue, inducing the tingling feeling—which, in itself, is not dangerous. The danger arises when neighboring, more significant, nerves are blocked, leading to paralysis and depressed breathing. I do not think I need explain any further. It is self-evident that this culinary activity is a rather risky one—so much so that if you want to become a fugu chef in Japan, you must go through a formal educational program to obtain a license. According to several sources (although there is some controversy about these), the emperor of Japan is forbidden to eat fugu because of its potentially lethal qualities.

Tetrodotoxin is not found solely in fish, however; we find tetrodotoxin in a diverse series of animals. For example, TTX is also present in the blue-ringed octopus, in certain species of marine and terrestrial flatworms, in a few kinds of poison arrow frogs, and even in several types of newts.

BEWARE OF THE LIVING DEAD

Tetrodotoxin and related molecules may be the source of a fascinating cultural legend, namely the concept of *zombies*, one of the most popular icons of the horror genre. I do not think that I have to explain to you what zombies are. What you may not know is the possible origins of the legend. Part of the lore of the voodoo religion, a set of beliefs predominant in parts of the Caribbean, is the ability of certain initiates to create "living dead," which ostensibly means dead people who come back to life, albeit under the control of others, these others being, of course, voodoo practitioners. The first part of this practice is to "kill" a person with a mixture called "zombie powder," which supposedly induces a marked reduction in all metabolic activities, rendering the affected person, for all intents and purposes, dead. The possible connection between zombies and tetrodotoxin was explored in Dr. Wade Davis's 1985 book *The Serpent and the Rainbow*, later adapted into a movie.[7] TTX seems to be one of the main components of the zombie powder that plays a ceremonial role in some voodoo religious practices. However, in 2008, an article published in the magazine *Skeptical Enquirer* challenged Davis's interpretations.[8] Whether or not the zombie-TTX connection is entirely accurate, with fictional accounts of zombies more popular than ever, it remains an interesting link between very real toxins and this cultural phenomenon.

One of the most fascinating facts related to biological arms races, particularly in the case of toxin evolution, is that oftentimes predators and prey coevolve at the molecular level in the sense that toxins coevolve with their molecular targets. One of the earliest and still very interesting examples is the case of the common garter snake (*Thamnophis sirtalis* and related species) vs. the common California newt (*Taricha granulosa*).

In the fierce battle between these two creatures, the main biochemical protagonist is our old friend tetrodotoxin. TTX is the likeliest source of a semi-legendary account of newts and mysterious death that led to research illuminating this particular evolutionary arms race. According to the story, in the 1950s, three buddies went hunting and were never heard from again. After some time, the police found their bodies at their campsite. The cause of death was not evident and there was no indication of foul play. The only odd thing found at the scene was a pot of coffee with a very much boiled newt inside it. This unfortunate newt had probably been scooped into the pot when one of the men collected water from a stream to make the coffee.[9] In the 1960s, this story inspired a young student, Edmund D. Brodie Jr., who began to study the toxicity of the California newt at the suggestion of his professor, Dr. Kenneth Walker. Brodie was indeed amazed at the lethality of newt skin extracts injected into a variety of subjects including mice, fish, and birds. Sometimes the extract killed mice within minutes of injection! While the fate of lab animals is a sensitive topic today, at the time this outcome encouraged young Brodie to continue. Alas, to Brodie's dismay, while he was performing these experiments other scientists discovered (and published papers on) the presence of tetrodotoxin in the very newts he was studying. These scientists also established that it was TTX that accounted for the newts' toxic properties. However, that did not stop young Brodie. There was an obvious question remaining to be asked, namely: Why do these newts need such a potent toxin to begin with? Brodie observed that many garter snakes fed happily on California newts and showed no ill effects whatsoever. In fact, these garter snakes seemed to be the only organisms capable of eating a toxic newt without dying. The evident resistance of some garter snakes to the toxic newts immediately suggested to Brodie that there might be a coevolutionary explanation.

Over the years, Brodie and his son, Dr. Edmund D. Brodie III, alongside other collaborators, unraveled this mystery. What they discovered was indeed a true evolutionary arms race, in which the resistance developed by the snakes exerted evolutionary pressure for the newts to develop higher toxicity, expressed as higher amounts of TTX toxin in the newts. This toxicity increase, in turn, induced even *more* resistance in the snakes. Over time this created a subset of the California newt that was more poisonous than ever, along with garter snakes that are incredibly resistant to tetrodotoxin.

Keep in mind that a single organism does not "evolve" in the sense that one fine day a certain newt decided that it was not poisonous enough to be safe from a garter snake and proceeded to make more tetrodotoxin at will. The same goes for the snake; it cannot "decide" to become more resistant to the toxin. So, how do they change? The key concept here is that individual organisms do not evolve by themselves. Evolution is a numbers game, based on those individuals that survive in a population under a particular set of environmental conditions. In other words, populations evolve; single organisms do not. This is the way it works, according to the natural selection interpretation of evolutionary theory.

The likely effect of this scenario will be "better" genes surviving in a particular population in order to cope with a given environmental change. Within limits, of course; genes cannot solve everything. For example, the aforementioned dinosaurs were probably the most successful big vertebrates ever, inhabiting this planet for hundreds of millions of years, only to become extinct because of what was essentially a big rock about six miles long, which upon falling to Earth approximately sixty-five million years ago, quickly and lethally disrupted the global environment. It is pretty clear that favorable genetics and fortuitous mutations are not enough to protect any organism from such a calamity![10]

EVOLVE IF YOU WANT TO LIVE

All of the aforementioned aspects of evolutionary arms races are an implicit reflection of the survival instinct that is pervasive throughout nature. When considering the topic of arms races and their relationship to natural selection, we have seen that there are some general steps—or a kind of algorithm, if you will—that result in the stepwise development of deadlier weapons and more efficient lines of defense against such weapons. What do these steps look like in the example of the newt-snake relationship? Well, let's suppose that you have a population of newts that is diverse in terms of how much tetrodotoxin a given individual newt produces. One of the environmental changes that challenges the newts comes in the form of the appearance of a predator, namely the aforementioned garter snakes. For example, a snake snatches one unlucky "low-TTX" newt, which is rapidly devoured and therefore prevented from passing its genes on to the next generation. Then the snake, still hungry, snaps up another newt, which just so happens to have more TTX in its system. In this case, the snake rejects it (and by the way, it looks like these newts indeed taste terrible to the snake, which "gags" and everything). Now, this second newt, having survived its ordeal, has a higher chance of mating and will probably leave descendants that will likely inherit the higher tetrodotoxin-producing ability. Thus, the next generation will display a higher proportion of ever-so-slightly more toxic newts. On average, the whole population will be more toxic and therefore more able to survive a snake's attack. Then it is the snake's "turn" to evolve (although please keep in mind that all these events happen more or less at the same time). Suppose that a hunting snake just happens to be more resistant to TTX than other snakes. This means that it will be more likely to enjoy a hearty meal, as it will not need to spit out any of the toxic newts. A well-fed snake has a better chance of surviving and reproducing. Over time, on average, the snake population becomes more resistant to the toxin.

I am sure that you see where this is going. In all likelihood there will be several cycles like this, which may result in "super-toxic" newts and, in turn, "super-resistant" snakes. In fact, scientists have observed these changes at the molecular level. The garter snakes seem to be the current winner in this war. Apparently, this is due to the relative sizes of these adversaries. Newts tend to be small, and a single newt can have only so much toxin, while the most resistant snakes, significantly bigger than the newts, can tolerate an amount of toxin close to ten times higher than the maximum amount of TTX in a single newt. However, life goes on and keeps evolving as it does. The end of this story is not yet in sight, and I would not be surprised to hear of further escalations in this particular tale of biological warfare.[11]

An interesting spinoff of this story features a type of insect generically known as the caddisflies. These insects, which are more closely related to butterflies than to proper flies, have larvae that like to feed on, among other things, TTX-laced newt eggs. Recent research seems to indicate that the caddisflies' larvae acquired their TTX resistance in a way similar to that by which garter snakes acquired theirs, only with results more extreme. For example, scientists estimate that resistant snakes can tolerate tetrodotoxin amounts about thirty times higher than can non-resistant snakes. On the other hand, caddisflies are capable of tolerating TTX concentrations two hundred times higher than non-resistant snakes.[12] Some caddisfly larvae are not only resistant to TTX-laced newt eggs, they even eat the eggs and then sequester the tetrodotoxin in their own bodies, presumably to be used for defensive purposes. In this case, obviously, size is not the determining factor; it must be a matter of their biochemistry. Tough snakes, toxic newts, and even tougher bugs—all courtesy of a natural toxin working alongside evolutionary changes. These are just a few of the intriguing strange survivors we'll meet on our journey through this book. Let's keep going, shall we?

CHAPTER 2

THE LANGUAGE OF LIFE*

B iological evolution, as important as it is, only works within the context of established life. In turn, life cannot possibly occur without its underlying chemical principles. At the most basic level, living organisms are exquisite, almost incomprehensibly complex contraptions, with an unfathomable number of interacting parts. Our understanding of the molecular mechanisms that make life possible is rudimentary at best, but one of the greatest hopes we have of increasing this understanding is by applying chemical principles, some of which we'll go over—very briefly—in the next section. However, if you are pretty comfortable with chemistry, or even if you simply find yourself unable to wait to learn about other curious organisms, feel free to go straight ahead to the next chapter. It will be our little secret. Otherwise, stick around, as we talk a bit more about life.

---

* With apologies to all other authors who have used this phrase in the scientific and popular literature, including book titles. The truth is that I cannot think of another phrase that better expresses the spirit of this section.

WHAT IS LIFE?

This is, of course, a trick question. Life is one of those enigmatic words that can mean different things to different people. Many people have tried to define "life" without much success, but here is one of my favorite definitions, courtesy of Democritus, a famous philosopher from ancient Greece: *"Life is the result of a special combination of atoms characterized by their constant mechanical movement."* Please note that Democritus was born around 460 BC—not bad for a guy who lived about 2,500 years ago!

As you would expect, in this book I use the word "life" in its biological sense. You may think that when we use the term exclusively to apply to biological life we'll have an easier time pinning down just what it is. After all, every known example of the living world is a physical, concrete entity. It is not like we are talking about love, meaning itself, or other such nebulous or contentious concepts, right?

Well, no such luck. The fact is that entire books have been written trying to explain what life is, with rather limited success. Over the years, some of these books sparked the imaginations of young people, inspiring them to become scientists, who in turn tried to find answers to life's mysteries. One such volume is widely credited with inspiring the molecular biology revolution of the twentieth century. I am talking, of course, about the 1944 book aptly titled *What is Life?* by Erwin Schrödinger, a physicist and Nobelist famous for his work on quantum mechanics. Schrödinger's is not the only book with that title. There are a few others, some really good, some really bad, and some really ugly. In any case, no universally accepted definition of life came out of any of those books. Moreover, there are even books that list more than one formal definition of life; as far as I can tell, the book that lists the most gives no less than forty-eight distinct definitions!*

---

* The book in question is *Biogenesis: Theories of Life's Origin* by Professor Noam Lahav of the Hebrew University of Jerusalem.

In great part because of the difficulty of defining "life," I will not fall into that trap. I will not even try to define what life is, much less drag you along that path with me, so don't keep reading hoping that a profound definition will pop up. The truth is that no one knows how to define "biological life" in such a way that everyone will be completely satisfied.*

The good news is that even though we do not know quite how to define it, we can competently describe what life does and what it is made of. So, what makes life tick? Simply put, life works by integrating the following four factors: chemistry, evolution, energy transfer, and information transfer. We can further simplify these four features of life, because any energy and information transfer activity within a living organism is a matter of chemistry. Thus, we can express the physical meaning of life using just two words: "evolution," which we have discussed in some detail already, and "chemistry."

There's no denying that life obeys the laws of physics and chemistry. Anything and everything alive is made from exactly the same set of chemicals. There is no such thing as a "life-exclusive" atom. Yet, a collection of many different atoms, when organized in specific ways, somehow generates the phenomenon of life, pretty much like Democritus said. Again, we still do not understand *how* this happens. I am confident that science will figure it out eventually, but I have a hunch that we still have a long way to go.[13] While scientists work on how life came to be, let me offer you five undeniable facts about this remarkable phenomenon: (1) there is life on Earth; (2) all life is essentially made of the same components, especially carbon**; (3) life on Earth is very complex; (4) absolutely every living thing on

---

* By the way, from now on, and unless I say otherwise, whenever I say "life" I will mean "biological life." Agreed?
** Carbon, oh sweet, mysterious carbon! It is only the fourth most abundant element in the universe, but it does not even make the Earth's top ten! And yet, all life that we have observed so far is carbon based, no exceptions. Moreover, many thinkers believe for very good reasons that if and when we find life elsewhere in the universe, it will most likely be carbon based as well.

Earth is related to one another; (5) and finally, every single type of organism possesses a built-in tendency to stay alive to the point of reproduction (the aforementioned survival instinct).

LIFE IS ORGANIZED AND COMPLEX

We are all a mixture of chemicals, but we are not just lumps of chemicals indiscriminately mixed together. From very simple molecules all the way to highly complex structures, living organisms exhibit a high degree of organization (the very root of the word "organism" refers to their organized nature). Everybody knows that atoms and molecules make up every single speck of matter in this universe. This fact also applies to us, since as part of this universe, humans are also made out of atoms and molecules. However, it is important to point out a very curious fact. No single isolated atom or molecule is alive, not even the ones present in living organisms. Nope, not even a little bit. Yet life is undeniably a chemical phenomenon, there's no question about it. Pretty much wherever we look, there is a very clear distinction between inanimate and living matter, and we have no problem telling one from the other. The difficult question is: Where is the boundary between living and nonliving matter? Paraphrasing the distinguished biologist J. B. S. Haldane's famous quote, if the fundamental unit of life is the cell, and since atoms and molecules are not alive, the boundary between live and dead matter must lie *between* the atom and the cell. Simple enough, right? You'd think so, but it is quite difficult to determine the exact point at which pure chemistry ends and proper life begins. No one has been successful at it so far. Furthermore, no one can even say that we have pretty good idea of where this frontier lies.

Then there is the small matter of viruses (pun absolutely intended). Most scientists do not count viruses as living organisms for a variety of reasons, one being that typical viruses are absolutely inert until they come in contact with an appropriate target cell. That said, all viruses possess genetic material, they can evolve, and when hijacking the host's cell machinery, they are able to make copies of themselves, which is without a doubt a form of reproduction. Also, it is undeniable that viruses have the capacity to interact with life at an intimate level. For example, we know of many viral species that can make humans sick, from the common cold to the terrifying Ebola. In light of all this, there are some schools of thought that consider viruses a link between living and inanimate matter. No biologist denies that viruses are quite mysterious in their own right![14] Yet despite all their lifelike characteristics, viruses are not quite alive. This is, of course, highly debated, but we'll stick with noncontroversial forms of life for the purposes of this book.

IF YOU WANT LIFE, YOU'LL NEED AT LEAST ONE CELL

What is a cell, anyway? You've probably heard the cell defined as the fundamental unit of life. Another good definition of a cell is the smallest self-contained collection of chemical compounds that, when working with other cells in very specific ways, possesses the capacity to somehow capture energy from the environment and use that energy to subsist and eventually make copies of itself. In a nutshell, no cells, no life. Thus, any individual living being is composed of at least one cell—and oftentimes many. *Unicellular* organisms are the most common examples of life, but because they are formed by just one cell, the common consensus is that their unicellular nature limits their relative complexity. However, simplicity can be deceiving. To be simple is not necessarily to be defenseless. Think about bacteria, some of which can be fatal to humans. In fact, unicellular life represents the most varied and successful kind of life. You may be surprised to learn that quite a few types of unicellular organisms engage in cooperative behaviors (we'll talk more about this in Chapter 6). Organisms like bacteria were here way before us, and they will likely still be around once our own species evolves into some other thing (at best) or (at worst) becomes extinct.

In a *multicellular* organism there are various types of cells that perform specific functions and cooperate with each other to try to ensure survival and reproduction—which we know comprises the primary mission of life as far as biology is concerned. A cell that survives and reproduces is successful in its mission by definition since it is assuring the continuation of its lineage.

Science textbooks occasionally represent the cell structure and organization in a rather simplistic fashion. I understand the practicality of doing this, because by its nature a textbook should express simplicity and clarity, but it is prudent to acknowledge the complexity of the cell. Cells are molecularly crowded, with a multiplicity of macromolecules buzzing around in apparent chaos, yet somehow, with very few exceptions, these subcellular components are able to interact correctly with their targets, keeping everything in order. Also, cells, like everything else in the world, exist in a three-dimensional space. Disregarding the tridimensionality of life is akin to saying that a stick figure drawn on a piece of paper is equivalent to an actual human.

One of the most important requirements for considering a group of chemicals as components of a "living" system is their relative, yet not absolute, isolation from their external environment. Cells are units that must therefore be separate and distinct from each other as well as from their external surroundings. Cells are able to do this because they have a flexible yet resilient and rather versatile barrier that we know as the cell membrane. This membrane, despite delimiting the cell's boundaries, nonetheless allows for the selective passage of chemicals through it. Just like a security gate, the cell membrane allows some substances to pass but not others. Some of these chemicals are informational molecules, some are nourishing molecules, and yet others are noxious molecules mostly derived from waste products, on their way out of the cell. I will not elaborate here on the particular structures that are found inside the cell, but there are a number of good references out there that you can go to if you wish to explore this interesting topic further.[15]

I AM A MUTANT, AND SO ARE YOU

"Just who are you calling a mutant?" you might very well be asking. Mutations have gotten a bad rap, because in the vernacular, we usually associate them with negative characteristics. However, mutations are nothing other than variations on the DNA sequence, and since we can agree that, with few exceptions, every single person on Earth will display slightly different DNA sequences, in a literal sense *everyone* is a mutant, and yes, that includes you (and me).

We can go further than that to say that not even your identical twin would be exactly like you. This is so because your heredity is not the whole story. How the environment interacts with your genetics also counts . . . a lot. Furthermore, several types of heredity exist within the same organism. Many of these are not completely understood. This is a fascinating topic for another day,[16] but here let's take a look at just how mind-boggling the genetic "facts of life" truly are.

All living beings react to their environment in various ways. We call the many ways in which an organism reacts to the external world its *behavior*. All life displays behavior; this characteristic is essential to perform activities related to nourishment, protection, or reproduction. While the usual definition of behavior is heavily biased in favor of animals, plants unquestionably also display behavior. We can truly say that any given organism will display behavior at different levels, from its individual cells all the way to the interactions between separate organisms and even between species. This is an essential aspect of biology; without some kind of behavior, living things would be indistinguishable from say, rocks. Behavior is, of course, also essential for reproduction. In the ongoing effort to survive, all known life makes use of captured energy to produce others of its kind through some type of reproductive process. Several varieties of reproduction strategies exist in the natural world, but we'll not talk about them here, as I assume that everyone reading this knows about the legendary birds and bees. Rather, let's talk about how life is able to pass its characteristics from one generation to the next regardless of the mechanical specifics of their reproductive practices.

Everybody nowadays is familiar with the acronym "DNA," even if what immediately comes to mind are police or science-fiction TV shows. Most people even know that DNA (which stands for deoxyribonucleic acid) is somehow responsible for what we look like and how we operate. Some people also have some notion that DNA somehow influences our sensitivity to pain, or the propensity to alcoholism in some people, whether we are prone to anxiety or depression, or even how we may react to a variety of life situations, among many other traits.

Well, DNA does none of these things, whatsoever . . . at least not directly. Surprised?

You see, DNA is essentially an information storage molecule, nothing more, nothing less. By itself, DNA hardly does anything. For DNA to "work" (in other words, for DNA to control every aspect of the making of an organism, from molecules to behavior), the information it contains needs to be extracted and then translated into a useful form capable of actually doing the things that life does. Specific sequences of DNA can be loosely likened to a barcode, which upon decoding can give us a lot of information about the specific type of organism from which it came. Despite the close similarities between the genetic material in organisms of the same species, interestingly, no one belonging to your same species in the world has *exactly* the same DNA sequences as yours, unless you have an identical twin— and even then, as mentioned, despite having the same genes, identical twins can display different tastes, birthmarks, opinions, and even different disease propensities!* In all likelihood, there has never, ever been anyone genetically identical to you. How do we know this fact? Just look around at other people (but don't stare!).

THE ONE AND ONLY YOU

People who know what they are doing and have thought about these matters reliably estimate that in all of human history, about one hundred billion people have lived and died on this wonderful planet of ours. Since the current human population is about seven billion, this means that as far as we know, in all the history of the universe, there have only been about 107 billion humans in existence.[17] With very limited exceptions, each of these one hundred or so billion humans has been genetically unique. It is very common nowadays to hear people stating that each of us is a unique and unrepeatable individual; we often hear this within a motivational context. Well, it is true enough, no question about that. If we know some biology, we can see just how literally true this statement is. Let's briefly calculate how improbable you are.

After puberty most women have—on average—close to 3,000 eggs that can potentially be fertilized to make a baby.[18] Typically, a woman ovulates 300 to 400 eggs in her lifetime, and under normal circumstances, only one is "released" per month. Based on these biological facts, any particular egg from your mom had a probability of about 1/3,000 to be available to make a baby

---

* You need to go no further than the many science fiction stories with clones as the main protagonists. Take for example the 2002 movie *Star Trek: Nemesis*. Briefly, some bad guys get ahold of Captain Picard's DNA and create a clone of him, named Shinzon. However, despite having the exact same genetic material, Shinzon and Picard grew up in drastically different environments. They ended up developing different personalities and values yet possessed equal intellects, which made them the perfect mutual antagonists, a fact that drives the plot of the story (watch the movie, it is pretty good!).

each month.[19] When you were conceived, an average of 200 million sperm from your dad competed to be the one that fertilized your mom's egg. Therefore, "sperm-wise," any given sperm cell had about a 1/200,000,000 chance of fertilizing an egg.

With just these two numbers, we can estimate a preliminary probability of you: 1/3,000 x 1/200,000,000 = 1/600,000,000,000. That's right, you are not merely one in a million; you are one in 600 billion!

In fact, there are more possible genetic combinations than all the humans that have ever lived or that will ever live in the foreseeable future. It is simply mind-boggling that DNA, a deceptively simple molecule, is capable of generating such variation. Moreover, probably the most astonishing thing about the genetics on this planet is that there are not multiple "kinds" of DNA; there is only one class of it. Therefore, there is only one type of life on Earth and it is DNA-based life. How is it possible that exactly the same type of DNA can generate not only the immense array of possible outcomes for a single individual but also the incredible variety of life forms in both type and sheer numbers on this planet of ours? Spoiler alert: It's mostly a matter of the sequence. The enormous variability of DNA, which codes for all kinds of life, comes from the way information is encrypted into a specific sequence. Only four basic building blocks, present in various amounts and in different order, are generally more than enough to account for the majesty of life around us.

Normally, DNA is organized within cells in one or more *chromosomes*. Prokaryotes and archaea (two types of single-celled organisms) usually have just one chromosome, and eukaryotes (usually multicellular organisms like plants, elephants, tigers, and you) tend to have many. Each chromosome in turn contains many *genes*. As you probably know, a gene is the so-called unit of heredity, which, in a very simplified way, can be defined as a certain DNA sequence within a chromosome that codes for a particular protein.[20]

To readily understand the survival concepts that underlie this discussion, let's quickly become familiar with three key terms: genome, genotype, and phenotype.

Genome: The genome of any organism is the summary, as it were, of all the genes it contains. Each species has a characteristic genome, immediately recognizable as unique. There is a rice genome, a fruit fly genome, and of course, the human genome, among many others. Humans, for example, have some 20,000 genes or so, at least at the most recent count,[21] and we are not even the species with the most genes, not by a long shot! For example, a tiny species of crustacean, a water flea called *Daphnia*—barely bigger than this letter *o*—has about 31,000 genes. How can 20,000 genes make an organism arguably more complex than another one with 31,000 genes? This gives rise to an interesting and very important fact. In general, the number of genes or the amount of DNA in the cells of an organism has virtually no relationship with its complexity. For example, if we could line up and stretch the DNA contained in a typical human cell, we would measure it at about six feet long. In the case of humans, cells pack 46 of these long DNA ribbons in a volume close to 500 times smaller than the period ending this sentence. Do you want to be even more amazed? If we line up and stretch the DNA within a cell of an otherwise unremarkable, although rare, plant called *Paris japonica*, we would end up with a DNA ribbon more than 300 feet long! Without any doubt, there is still a lot to wonder about in fundamental biology.

Genotype: As we said before, DNA does next to nothing (directly, that is). When we say that DNA is "active," it mainly means that a particular DNA sequence is read and eventually that information is used by the cell to make at least one protein. In turn, proteins are the "muscle" behind DNA's "brains." Simply put, the expression of proteins largely determines what a particular organism looks like and what it does. The genotype is simply the genetic instructions that are encoded by the genes.

Phenotype: Our third term, the phenotype, can be defined as the expression of the genetic information resulting from the genotype, or genetic instructions, operating in tandem with environmental influences to direct cellular activities. Common examples of phenotypes include traits that are immediately apparent, such as hair and eye color, body shape, agility, etc. Other aspects of the phenotype may include traits that are not possible to notice just by looking at an organism. These include how an organism will react to a particular medication, if an organism is allergic to something, or even if an individual is prone to mental diseases. In other words, the genes (the genotype) and the environment collaborate to express every single characteristic of the physical reality of all organisms (the phenotype).

Here's one important takeaway: Each and every one of the curious survival mechanisms that we'll see throughout this book are phenotypes.[22] Yes, the majestic glide of an eagle, the delicate colors of a rose, the aggressiveness of a praying mantis, the rotten smell of the corpse flower, and our own mental capacities—all things good, bad, and ugly that an organism can do are examples of phenotypes.

LIFE'S WORKHORSES

DNA is sometimes called the language of life,* which offers us a useful way of understanding how it contributes to the survival instinct all living organisms share. Thus, it might be useful to touch on what a cell *does* with the information encoded in DNA. One of the fundamental functions of DNA is to make proteins, and proteins do pretty much everything inside a cell. Proteins are the immediate product of this so-called language of life. This language is organized into what

---

* Hence the title of this chapter . . .

THE LANGUAGE OF LIFE

we call the *genetic code*. Proteins include some of the more complex molecules in biological systems. I will not go into too much detail about them, as other works have covered this topic masterfully.[23] One very interesting point is that life uses only twenty specific amino acids to make proteins. However, there are some five hundred described amino acids found in nature. What is so special about "the twenty"? Nobody really knows. At the time of this writing, there does not seem to be any evident structural feature of the amino acids in proteins that is fundamentally different from all other amino acids found in nature, or any special quality that makes them especially suited or particularly advantageous for life. In fact, a few years ago, a couple of research groups developed a series of techniques to incorporate unnatural amino acids into relatively simple microorganisms via genetic engineering. These organisms seemed to display normal physiology and behavior. The reasons behind life's selection of the specific twenty amino acids that it uses presents a major biological mystery worthy of attention. I wouldn't be the least bit surprised if someone is working on that biological problem right now.

Most proteins include a series of some of the most interesting molecules that are responsible for what a cell does, in other words, its physiology.* We know these molecules as *enzymes*. Enzymes are intimately related to the chemistry of life. They are largely responsible for all the multiple chemical reactions that are necessary for an organism to pursue its ongoing mission—to stay alive.

EVERYTHING IS HUNGRY

The capture and use of energy is a nonnegotiable characteristic of life. We can classify any organism on our planet based on what it does to get energy (nourishment) from the environment—in other words,

---

* Physiology is essentially about the internal functioning of the cell, what we'd call *behavior* in an independent organism. This is a rather simplified statement, but it is enough to keep us going!

what it eats. Claiming energy is probably the very first thing that any being must do to survive; it is not optional.

Photosynthesizing plants and a few other organisms, for example, make use of their amazing biochemical machinery to capture radiant energy in the form of light. Along the way, they proceed to make the molecules that they need for sustenance. We call organisms—such as green plants—that are able to produce the necessary molecules for survival by capturing energy directly from the environment *autotrophs* (self-feeders). On the other hand, any organisms that need to eat other organisms for survival—even if it is just a salad—are classified as *heterotrophs* (those that feed on others; we humans belong in this category). This book is about heterotrophs.

There are many types of molecules that organisms require to survive and thrive. Right on top of the proverbial wish list is *adenosine triphosphate*, or ATP for short, which is the universal bioenergetic molecule. Every living being, from the smallest bacterium to the biggest whale, uses ATP to survive. We could almost say that in a literal sense, survival is the quest for ATP. But, we do not eat ATP, not directly; rather, we make it by taking the nutrients that we ingest and using part of their chemical energy to synthesize ATP.

Based on the very definition of what a cell is, it is self-evident that energy is an indispensable requirement for life. In fact, living in the biological sense is an endless search for energy. Energy use is such a defining feature of life that some authors have amended the famous quote from Theodosius Dobzhansky, "*Nothing in biology makes sense except in the light of evolution*"[24] by further stating that "*nothing in evolution makes sense except in the light of energetics.*"[25]

HOW MUCH DO YOU MAKE?

Do you want to know how much ATP you make per day? I guarantee you will be surprised. In any living being, ATP is being continually synthesized and used, and you are no exception. Without going into too much detail, we can estimate how much ATP an average human body makes per day based on a daily consumption of about 2,000 calories, a normal dietary intake. To make a long story short, you and I make—and immediately use—close to 46 kg (about 101 pounds) per day. Yes, per day! Yes, I said 101 pounds. Yes, I double-checked the calculations.[26]

LIFE IS BOTH DYNAMIC AND CONSTANT

Another fundamental characteristic of life is that it is not a static phenomenon. A living organism is in a constant state of movement, particularly at the molecular level. This constant state is intimately related to what we call *metabolism*, which is the summary of all the chemical reactions in an organism. Many of these reactions are related to the process of self-maintenance that is ever present when an organism is alive and are frequently related to structural and energy-producing reactions. Each and every example of an organism ever studied in nature undergoes constant self-repair. This means that there is a constant flow of energy and matter in and out of the organism at any given time. Oftentimes we hear that every year close to 98 percent of all the atoms in our bodies are replaced, and this is not very far from the truth. This is another fundamental property of biology that surely tells us a lot about what life may be, exactly. Alas, the replacement process is not always perfect; if it were, there would be no such thing as aging.

The essential molecular movement activities that guide life's self-maintenance are collectively called *autopoiesis*. As far as we know (and we have looked!), only proper cells are capable of *autopoiesis*, and therefore it seems to be a primary property of life, perhaps the *original* trait that allowed a mixture of chemicals to be truly alive.[27]

Modern biology views metabolism as an indispensable characteristic of life; there is really no alternative. As such, metabolism is exquisitely regulated in multiple ways. One of my favorite ways in which scholars describe such regulation is by comparing the chemistry of life to a juggler.[28] A juggler must not be too stable. If she or he is too stable, there will be no movement, and if the juggler does not move, everything that is up in the air will fall outright. Conversely, if the juggler moves too fast, it is very likely that she or he will lose control of what's up in the air, with potentially disastrous results (especially when juggling chainsaws or knives, for example). Just like a good juggler, life must maintain a well-regulated equilibrium between motion and stillness. Now, in my view, this equilibrium may hold the true secret of life, more so than DNA, reproduction, or perhaps even evolution! But only time and more research will tell.

THE ORIGIN OF LIFE

Sadly, this is one of many mysteries that we have little hope of ever solving completely. This key moment happened about three and a half billion years ago,* giving geology plenty of time to erase any evidence that might tell us how life got its start. And this is assuming that life started here on Earth! Serious scientists have raised the possibility of life being truly *extraterrestrial*, meaning that it developed elsewhere, and eventually "infected" our planet. However, even if this turns out to be true, it is does not answer the question of how exactly inanimate matter makes the transition to animate matter. Plus, again, if life indeed came from "out there," it happened billions of years ago . . .

Until very recently, we had very little idea of how life began, but it is a safe bet that chemistry played a central role in this story. For more than sixty years, we have known that some relatively ordinary compounds that are part of our current biology can be easily formed in the absence of life. It seems simply to be a matter of mixing raw

---

* Some recent data seems to indicate that this event took place about 3.8 billion years ago. The jury is still out on this one, and at any rate, it does not change our argument.

materials in the form of molecules like carbon dioxide, methane, ammonia, etc., with plenty of water, and adding some source of energy—such as heat and electric sparks, for example. When scientists conducted this kind of experiment, they found that in a rather short time, basic compounds like amino acids, the building blocks of proteins, began to appear in the water. This story began with the classic 1953 experiments of Stanley Miller and his advisor, the Nobel laureate Harold Urey, at the University of Chicago. Their work was a culmination of a line of thought that commenced when we humans first started to learn about chemistry and its relationship to life as we know it. Most accounts trace the origin of the scientific search for the origin of life to a series of scientific insights from the 1800s. Two of these were the scientific formalizations of chemistry and biology by the Russian chemist Dmitri Mendeleev and the British naturalist Charles Darwin (of course). Sometime later, two leading thinkers who proposed the idea of linking the chemical composition of the early Earth with the origin of life were Aleksandr Oparin (another Russian scientist) and J. B. S. Haldane (another British scientist whom I have mentioned), who in the 1920s independently came up with the idea of the "primordial soup," or what you may know as Darwin's "warm little pond," which in turn inspired the University of Chicago experiments mentioned previously. Based on the results of these experiments, it seems that the synthesis of biologically relevant compounds is a relatively simple matter. At the very least, Miller and Urey demonstrated that some of the molecules of life could have appeared in the proverbial "warm little pond" imagined by Darwin and later by Oparin and Haldane.[29] However, this is far from the whole story and these were not the only scientists who tried to tackle this problem. Now we think that life did *not* originate in any warm little pond, but at least in part in some hellish environments that exist at the bottom of the ocean, under conditions of very high temperature and pressure. There is always more than one side to any story. Probably, several scenarios were involved in the origin of life, even some that we may have not considered yet. Only time will tell how close we will come to the complete answer. There is still much to learn about how life came to be, and many authors have written about this rich and interesting topic in great detail.[30]

Before we close this section, it is crucial to establish an important point. Despite what some blockbuster movies might suggest, no living critter has ever crawled out of any experimenter's glassware. Clearly the origin of life takes much more than the indiscriminate synthesis and mixture of mere chemicals. These chemicals must organize in a particular way (or ways) under particular circumstances that we have not figured out . . . yet.

MYSTERIOUS DEATH

As I said at the onset, death is the rudder that gives directionality to the evolutionary process. Selective death brought forth by changes in the environment makes room for the critters able to survive those changes to reproduce. In purely physical terms, death essentially occurs when an organism reaches a state of equilibrium with its external environment. This simply means that there is no difference between the internal environment of the cell and the outside world. Think of a house in the middle of a harsh winter; if the heat is not on or if there is no furnace or chimney in working order, pretty soon the interior of the house will likely be as cold as the air outside, especially if the windows and doors are left open.

Death is truly an inevitable part of life, but it did not start out that way. Don't get me wrong. Any organism can be killed—and relatively easily too. This has been an inescapable fact of life since the beginning of time. But although death is inevitable, for early life it was "optional" to grow old and to die because of it. Biologically speaking, the inevitability of the so-called natural death is a relatively recent invention. There is every reason to think that modern microorganisms differ very little from their original forms. And the fact is that bacteria and related beings do not really grow old, mainly by virtue of their mode of reproduction. Simply stated, once a bacterium grows big enough, it splits into two new bacteria, and both merrily swim away until the next time they divide. As long as there is enough food available, this will keep happening. This state of affairs continued unchanged for a long time, but eventually something happened: "natural" death itself evolved, and cooperation between cells seems to have been the cause.

More exactly, and not surprisingly, the most likely origin of natural death was war, specifically biological arms races. In brief, as with virtually every biological trait, the invention of death seems to have been a consequence of the interaction between genetics and the environment. If we have a large population of cells, say bacteria, even though they may be genetically identical, they may not live in exactly the same environmental conditions. This means that different cells will likely be exposed to slightly different environments in terms of temperature, pressure, and light availability, among many other factors. In turn, the cells will respond differently. There is evidence that indicates this may be a form of cell specialization, oftentimes known, appropriately enough, as *differentiation*. We know all about differentiation in the context of multicellular organisms. For example, with the lone exception of your sex cells, all the cells in your body are genetically identical, yet a nerve cell looks and works very differently from a liver cell, for example. The phenomenon widely known as programmed cell death is a well-documented and established mechanism that is essential in most living organisms, and it is intimately related to differentiation. This programmed cell death likely started with unicellular bacteria, yet is nowadays an integral component of the formation of organs or appendages in multicellular organisms.[31]

To further illustrate the idea that natural death is biologically optional—up to a certain point, and at least at the organism level—there are several kinds of living beings that have done away with old age altogether. Yes, you read that right. Unlike humans, there are organisms that seem to have solved the problem of aging. Not only do they not seem to grow old, there are some organisms that can go back to their younger selves! We'll meet some of them as we continue our exploration of the extraordinary survival strategies nature has created, including truly exotic living beings that have learned how to use the natural electricity produced by their own bodies to enhance their chances of survival. In my mind, this capacity earns these creatures the distinction of being known as the planet's very first electrical engineers.

CHAPTER 3

IT ALL STARTS WITH A SPARK

The very first time humans saw electricity in action must have been a moment of indescribable beauty that almost immediately took a truly terrifying turn, especially if it happened at night. As with most of these "firsts" in the human journey, there is no historical record of this occurrence—it is very likely that people witnessed lightning and thunder well before it even occurred to us to document the events in our lives. However, we can easily imagine a group of ancient people settling down at night. Children were finally quiet, and everybody was about to retire after a long day of "hunter-ing" and gathering. They could not know that they would soon experience what is widely regarded as "one of the best free shows" that nature has to offer.

The splendor of lightning seen from afar, the kind of lightning that illuminates a significant portion of the sky, must have taken them by surprise. To the inquisitive mind of a primate—make no mistake, we humans are primates—this spectacle must have been a rather beautiful yet curious event, the kind of moment that inspired a sense of wonder. Then, a couple of seconds after the light, the sound arrived, and it was not a particularly pleasant one, as the sound of thunder is invariably ominous.[32] In this particular instance, let's say the storm was heading toward the tribe, meaning that as time passed, lightning and thunder occurred closer and closer together. Eventually, the light and the blast would occur simultaneously in a terrifying yet almost otherworldly beautiful display of brightness and sound.[33] Witnessing such an event would have likely kindled the curiosity of our ancestors, who eventually came up with many ideas to explain how lightning and thunder came to be. These early explanations were usually connected with spiritual or supernatural beliefs. Inspired by the raw power of this natural phenomenon, virtually every culture throughout history has worshipped weather gods, with Zeus and Thor being only two of the most familiar examples.

Just like with life, describing what electricity does is quite easy, but defining it precisely is pretty difficult, and this was how things were for a very long time. However, the fact that we humans do not know the nature of something has never stopped us from naming it. Take the very term *electricity*, for example, coined near 1600 by the British polymath William Gilbert, who is widely regarded as the "father of electricity." Ironically, Gilbert coined this term in a book about a seemingly different phenomenon, magnetism (his book was actually titled *On the Magnet*). He strongly argued in favor of the idea that electricity and magnetism were different phenomena and suggested experimental approaches to prove his ideas. Alas, it was not to be, as Gilbert was wrong. It turns out that at its very core, magnetism is another aspect of electricity—and vice versa. One cannot exist without the other. We know now that electricity and magnetism are two sides of the same coin, so to speak, hence the term *electromagnetism*, a fascinating scientific discipline pioneered by the likes of the Scottish scientist James Clerk Maxwell and the legendary Albert Einstein.

THE ORAL EXAM

There is a quite amusing and often retold—as well as most likely apocryphal—story about how little we know about the precise nature of electricity. This story supposedly happened at Oxford University close to one hundred years ago. Lightly paraphrased,[34] it tells the tale of a physics oral examination in which a student was asked what electricity was. The student was rather nervous* and replied something along the lines of: *"Electricity? I knew what it is, but I forgot . . ."* The chief examiner, sighing and most likely amused and slightly irritated at the same time, remarked something like, *"Pity, only two people have ever known what electricity is: yourself and the creator of the universe, and now one of them has forgotten!"*

One of the reasons electricity is so mysterious to us is that we use the term to describe a couple of types of phenomena that behave more or less in the same way. Most times we define electricity in terms of a description of the behavior of certain negatively charged particles (electrons, but electricity can also be carried by positively charged ions) and their movement from one place to another. Regardless of how we choose to define electricity, it is without dispute that our society runs on it in one way or another. Electricity powers up virtually all of the technologies that we associate with our modern world, but it goes far beyond that. You may be surprised to learn that electricity also runs the biological world. In fact, there is no biological process that does not relate to some aspect of electricity in one way or another.

---

* By the way, university oral examinations usually make students nervous; you can almost say that they are *designed* to be that way. I know this from personal experience as both a onetime student and now as a professor.

ELECTRICITY 101

Electricity comes in two main forms: static electricity and flow electricity. In either form, electricity is simply a type of energy that is part of the fundamental nature of matter. In a literal sense, we cannot say much beyond that. Depending on the specifics, we can encounter positive and negative charges, opposite in sign yet identical in magnitude. Negative charges are carried by subatomic particles called *electrons* (from the Greek word for "amber"), and positive charges are carried by another type of subatomic particle called *protons*.*

Flow electricity is just the measurable, more controlled movement of electrical charges, which is another of those terms that are not easily defined. It is enough for us to know that the property of electrical charge is one of those things in life that just "is." To summarize, we consider electrical flow the movement of charged particles, whether they are positive or negative. Flow electricity is involved directly or indirectly in every known biological process. In this book, we'll talk about electrical phenomena in living organisms, or *bioelectricity*.

NATURE'S TASERS**

Perhaps one of the most dramatic illustrations of how life uses electricity is the curious example of several species of fish collectively called *electric fish*. These organisms take advantage of their ability to

---

* And of course you know about *neutrons*, which carry no charge at all and therefore have very little to do with electricity.

** I wish I could take credit for the title of this section; it is a great one! I heard this expression for the first time at www.relativelyinteresting.com/meet-the-electric-eel-natures-taser/. However, I am a nitpicker by nature (I can't help it; most scientists and professors are nitpickers!). At the time of this writing, the website article displays a picture of a moray eel, which although quite scary, is not an electric fish. Also, as we'll see soon, the "taser" moniker as applied to electric fish recently appeared in a scientific publication.

produce significant "amounts" of electricity. They use that electricity to stun prey and enemies, and in fact, many call them nature's tasers,[35] so let's call them that too!

» ELECTRIC CATFISH

If you are even casually interested in natural history, you probably know about electric rays and electric eels. However, it may be a surprise to learn that there is such a thing as an electric catfish (*Malapterurus electricus*). It was certainly a surprise to me! I am a little embarrassed not to have known about electric catfish before researching this book, because I did my master's thesis on some aspects of the biochemistry of the electric organ in rays. Oddly, when the story of electric fish is told—even in scientific works—the electric catfish story is oftentimes overlooked.[36] This is strange to me; you see, unlike a ray, with its flattened shape and mysterious yet majestic glide, or an eel, with its snakelike appearance and demeanor, a catfish is, well, a fish, a fish that looks like a very common fish (please excuse my Dr. Seuss moment). I guess what I am trying to say is that I feel the electric catfish ought to be one of the more widely known "brands" of nature's tasers.

Nowadays, not many of us live in a place where strongly electric fish are ordinary enough to be frequently seen and perhaps be an inconvenience, or even a danger. However, a cursory examination of natural history records shows that our ancestors knew these types of catfish very well. There are many examples of illustrations in ancient Egyptian art dating from about 4,500 years ago of what are easily recognizable as the electric catfish. There are even stylized renditions of fish that might be considered portrayals of electric catfish in even older Egyptian art, these archeologically dated to at least a thousand years older. Early Arabic texts as well as early medical literature from other regions described electric catfish as capable of inducing "shaking." Aside from these written historical records, it is virtually certain that people knew of electric catfish long before they began documenting them. In fact, it is quite possible that people knew of them even before people were "people," meaning modern humans, as these fish have been around for at least twenty-five million years, and for all of those years have lived in Africa, the human ancestral homeland. There is no doubt in my mind that quite a few of our earliest ancestors must have felt this fish's shock at one point or another.

There are nineteen closely related species of electric catfish described in the current scientific literature. Electric catfish are widely distributed in African lakes and rivers, and one of the electric catfish's nicknames is "The electric fish of the Nile."[37] This nickname came to us from one of the great naturalists and explorers of his time and perhaps of all time: *Alexander von Humboldt*.[38] Incidentally, many types of catfish worldwide have the ability to detect electrical fields, but only *Malapterurus* catfish are electrogenic, meaning they are capable of producing significant electrical discharges. These guys can pack a punch too; *Malapterurus* can deliver an electric shock in the neighborhood of 350 volts! Sometimes this catfish was referred to as the *torporific* (sluggishness-causing) catfish, on account of the numbing effect a person felt when shocked. This is an important detail that, though people did not realize it at the time, hid significant clues about the electric nature of this fish. This fish's properties sparked the curiosity of a number of naturalists who made significant contributions to the discovery of electricity. Ironically, Europe learned about these fish for the first time not from naturalists, but rather from tradesmen, clerics, and settlers. Upon hearing stories and rumors of unusual beings, natural historians then proceeded to have their day studying them. Come to think about it, this was kind of the way things usually played out in the so-called Age of Wonder. Simply stated, word of mouth oftentimes led to scientific discoveries.

A SHOCKING STORY

I am very fond of saying that if there is anything in this universe capable of traveling faster than light, that thing is gossip. Here's a story that made the rounds in 1967. A news piece in the April 2nd edition of the *New York Times* reported that Dr. Frank Mandriota and his collaborators at Columbia University were able to train an electric catfish to trigger its electrical discharge by exposing the fish to certain stimuli like light pulses or even weak electric shocks. Essentially, they supposedly trained this catfish to "shock on command." This story is widely retold online, usually with the exact same wording on different websites, always a red flag when judging the reliability of information. If this story were true, it would have been quite exciting as well as rather dramatic (at least from the perspective of a biologist or even a fishing enthusiast). Imagine having a strongly electric fish that would shock other fish for you at your command! It would essentially act like a well-trained underwater hunting dog. I admit that this hypothetical scenario comes from my overactive imagination, but who knows? However, there are some aspects of the *NYT* story that cast doubt on the alleged facts. In the newspaper report, for which they seemed to have interviewed Dr. Mandriota, they stated that the researchers would publish their results within six months. Two years prior to the reported interview, Dr. Mandriota and his collaborators *did* publish a study about the training of an electric fish, but the experimental subject was not an electric catfish but a *mormyrid*, a weakly electric fish available at pet stores, a fish that we will talk about later in this chapter.[39]

Indeed, in 1968, the promised study from Dr. Mandriota was actually published, but again the animals studied were mormyrids, not electric catfish.[40] To date, no paper on electric catfish in this context has ever come to light from any research group. However, I was also unable to find a correction note in the newspaper or any type of clarification about this matter. Was this an instance of miscommunication? Was this a case of perhaps less-than-thorough fact-checking? Did the researchers actually repeat their experiments on a catfish and somehow never got around to publishing them? This latter situation does actually happen with some frequency in scientific research.

While this story can serve as a cautionary tale about taking everything you find on the Internet at face value, let's also make clear that this does not diminish the achievement of the Mandriota group one bit. They *did* train an electric fish and publish the research. In the eyes of a biologist, these results are as significant and novel as if the trained fish were indeed an electric catfish. We'll just not be able to use them as hunting (fishing?) fish, that's all. Bummer.

» ELECTRIC RAYS

Electric rays are a much more familiar example of strongly electric fish, and they are well represented in early natural history literature. Depending on the classification scheme, there are some seventy described species, including two (*Tetronarce cowleyi*, from southern Africa, and *Narcine baliensis*, from Southeast Asia) discovered as recently as 2015 and 2016, respectively.[41] Like electric catfish, electric rays were widely known by the ancients. As with most things in nature, they were the source of artistic inspiration as well as legends and stories. For example, it is evident to any casual natural history enthusiast that Greek philosophers pop up in pretty much every corner of the subject, and electric fish are no exception. Several of the most famous ancient scholars—Plato, Aristotle, and Galen, among many others—knew and wrote about what we know now as electric rays—most of the time they called them *torpedo*, a name that stuck. Electric rays are easily recognizable in many works of art, some of them rather poignant, such as certain colorful mosaics unearthed from under Pompeii's ashes. We can only imagine how beautiful that artwork was when originally created!

Among some early fishermen, electric rays earned a legendary reputation as noble protectors of smaller fish. This belief appears to have arisen from the fact that when fishermen pulled on their nets, if an electric ray was captured, it shocked or numbed the unsuspecting man, who promptly let go of the net, allowing the escape of all captured fish.[42] Even though this would not have been a happy occasion for the fishermen, the rays nonetheless earned the fishermen's respect for their perceived selfless act. There was, of course, no explicit intention from the rays to save their fellow fish—they were merely trying to defend themselves—and in fact, the rays would almost certainly eat some of these liberated fish given the opportunity.

In other places, people assigned these rays less benign characteristics. Sir John Pringle, a physician and naturalist as well as Benjamin Franklin's pal, studied electric rays quite extensively—and he found them fascinating. How could he not? However, you wouldn't think so from the way he talked about them, because Pringle famously wrote that the electric ray was "*a mean and groveling animal armed with lightning . . .*"[43] This, by the way, is an example of what is almost an occupational hazard for the nature lover: the personification of nonhuman organisms. To be fair, you would also get quite mean if you were poked, provoked, or otherwise disturbed, don't you think? Oh, more than once in my life I had wished to be electric so I could zap the . . . I shall mind my manners and tell you more about the fish.

ANCIENT MEDICINAL USES OF ELECTRIC FISH

Physicians of long ago listed a few diseases and conditions that were to be treated in one way or another by putting the patient in contact with electric rays. These medical conditions included general pain, migraines, and gout, among others. Here is a prescription to treat gout-induced pain, recorded by the early second century Roman physician Scribonius Largus:[44]

"For any type of gout a live black torpedo should, when the pain begins, be placed under the feet. The patient must stand on a moist shore washed by the sea and he should stay like this until his whole foot and leg up to the knee is numb. This takes away present pain and prevents pain from coming on if it has not already arisen."

Like electric catfish, electric rays were oftentimes described as being "torporific." More than a few people noticed numbing, tingling, and twitching upon coming into close contact with a variety of electric ray species. Sometime after Largus, an early physiologist, Stefano Lorenzini, discovered that the electric organ of torpedo rays emitted "something"—we now know it to be electricity—that could cause painful sensations or numbing when touched: we now know these as electric shocks. Even Galvani and Volta, famous adversaries who were the unlikely co-parents of the discovery of electricity in living organisms, agreed on the electric nature of the *torpedo* rays.

The use of electric fish to alleviate pain is not as far-fetched as it sounds. This practice was a true antecedent of some of our current physical therapy treatments. For instance, a common modern technique that physical therapists use to treat muscle spasms or pinched nerves involves applying heat, cold, and yes, even small electrical currents to the affected area. When properly administered, these therapies prove very effective for alleviating certain kinds of pain, albeit temporarily. Interestingly, certain electric ray species are named after their anesthetic-like effects. For example, scientific names of the aforementioned new species of electric rays, *Tetronarce cowleyi* and *Narcine baliensis*, contain the term *narce* (oftentimes translated from the Greek as *nárkē*, meaning "to numb"), in direct reference to these fish's anesthetic properties. These are by no means the only organisms that contain this word within their scientific names.

Whole genera of electric fish have also been named after *nárkē*. In turn, *narce* is the origin of several other terms widely used in modern times, all related to the loss of sensation or consciousness; think narcotic, narcolepsy, etc. Electric fish undoubtedly contributed to our health sciences before anesthesiology—or even medicine, for that matter—were formally established as scientific disciplines.

» ELECTRIC EELS

As important as electric catfish and electric rays are in the history of science, electric animals have one more surprise in store for us. Let's talk about the most powerful electric fish ever discovered: the electric eel. This strange creature is rather inaccurately named, because this taser is not an eel, not even close!

This organism in fact belongs to a group of creatures called knife fish, which are very interesting animals in their own right. The "eel" nickname is understandable though, as these are definitely serpentine animals. They may grow to be eight feet long and, as far as electricity is concerned, they take the indisputable first place, being able to deliver electric shocks in the range of about 800 volts!* There is only one recognized species of electric eel: *Electrophorus electricus*. Its stomping grounds are the northern and central parts of South America, but there are early reports indicating they were once found as far south as Argentina. These organisms use their strong electrical currents in various ways: to deliver the obvious shock, to track their prey, to induce fatigue in their prey, and to immobilize their prey for capture. Based on their observed behaviors, they seem to know their own capabilities.[45] Once the naturalists from the Age of Wonder

---

* Let's put the strongly electric fish's electrical charges in perspective. A normal electric plug in the Unites States is set to about 110 volts (220 volts for big electrical appliances, such as water heaters and the like).

learned about the various kinds of electric fish, they were indeed "struck with wonder" by them.

One of the early explorers who wanted to know more about this particular organism was Alexander von Humboldt, the very same person who named *Malapterurus* the "electric fish of the Nile." Von Humboldt's contributions to natural history were clearly above and beyond what many of his contemporaries ever achieved. He inspired a couple of generations of young naturalists, including a young Charles Robert Darwin, who years later would write a letter to Humboldt in admiration and gratitude for igniting his own dreams of exploring the natural world and its wonders.[46]

There is an interesting connection between von Humboldt and the electric eel. In 1800, von Humboldt wrote an account of an unorthodox technique that native South Americans used to capture electric eels.* The technique was colloquially known as "fishing with horses," the name being quite descriptive of the actual practice. Essentially, fishermen would force horses or mules into a body of water known to contain electric eels. Then they prevented the quadrupeds from leaving the water by screaming at them or hitting them, which of course agitated the big animals, making them splash and thrash in the water. The movement in turn distressed the electric eels,

---

* I love when, in researching one topic, I find an unexpected connection to seemingly unrelated subjects. For example, one of the earliest descriptions of the electric eel from South America came from a Spanish priest, Father Bernabé Cobo, who lived in Peru in the 1600s. He was a prolific natural historian, and in 1653 he wrote about a fish that he called the *"torpedo of the Indies."* He was of course referring to the electric eel; there are no other strongly electric fish in the Americas' freshwater systems, with the exception of the stray electric ray that sometimes comes from the ocean up by the river. I knew of father Cobo, albeit in a different context. He was the first person to write about the anesthetic properties of coca leaves (the source of cocaine). Briefly, he had a toothache, and his barber—in those days, barbers doubled up as dentists and surgeons—advised him to chew on coca leaves for pain relief. Father Cobo's report was probably one of the first inklings that indicated that coca leaves had something that served as an analgesic. The rest is, of course, history.

which then attacked the horses by leaping onto and shocking them. After a few discharges, the fish became tired and could be picked up safely. Along the way though, several horses would usually drown as a result of this fishing event. The practice was as barbaric as it sounds, and this account of events was doubted by subsequent generations of natural historians despite von Humboldt's reputation. These scholars fell short of accusing von Humboldt of lying, but they truly thought that the story was at best an exaggeration of the facts. It didn't help that the eels' alleged "leaping behavior" was not observed for close to two hundred years after these first descriptions. Then in 2015, Dr. Kenneth Catania, a biologist at Vanderbilt University in Tennessee, observed exactly such behavior when studying *Electrophorus* specimens. Upon using a net with a metallic rim and handle, the eels leapt out of the water and struggled to keep in contact with the net while discharging electricity.[47]

One of the main questions that people ask about strongly electric fish is: How do they prevent getting shocked by their own electrical currents? After all, they live in water, a highly conductive environment. This is an excellent question that for a long time remained unanswered, and this applies, of course, to both freshwater and marine species of electric fish. One of the main theories that tries to account for these fish's virtual immunity to strong electrical currents is that they have relatively thick layers of fat and connective tissue that help insulate their insides from their emitted electricity. However, in some cases, this protection is not quite sufficient, as there are observations of electric eels twitching in response to their own electrical discharges. These occurrences might be explained by an injury that caused scar tissue, which may leave "gaps" in the protective layer, which in turn may allow electricity leaks—much like a damaged power cable. There is ongoing research in this area at the time of this writing, but it is a safe assumption that some version of this mechanism is probably shared by all strongly electric fish.[48]

Regardless of the species, all electric fish produce electricity based on some kind of—aptly named—*electric organ*. In virtually every case, these electric organs are modified muscle tissue.[49] In contrast to normal muscle, electric organ tissue does not contract—or even move, for that matter. Upon close microscopic examination, it becomes apparent that the cells of these organs are stacked in a highly organized way. In modern times, the analogy between this cellular arrangement and the engineering of an electric battery became evident. Moreover, these cells contain a relatively high amount of a specific ion channel that controls the passage of small electrical currents throughout the cells. The combination of the following three aspects allows these organs to emit significant amounts of electricity: the intrinsic electrical nature of cells, the battery-like arrangement of the organ, and the high amount of this particular ion channel.[50] Hence, when activated, a significant amount of charged particles move, effectively expressing electricity. The story of how cells store electricity is amazing in itself. To help further understand how these glorious electric fish operate the way they do, I shall tell you that story now.

BIOELECTRICITY

The classical bioelectrical experiments using frog legs, originally done by Luigi Galvani and replicated by Alessandro Volta, paved the way for our understanding of the electrical nature of nerve and muscle function. The story of these two scientists, one-time colleagues by correspondence and more often than not scientific rivals, is well-known and rather well documented, so I refer you to excellent sources.[51]

But, what is bioelectricity, after all? In Chapter 2 we talked a little bit about the organization of the cell. At its most fundamental level, the cell as a unit of life is defined by the presence of an oily structure called the *cell membrane*, which surrounds all cell components and isolates its interior from the external environment. This is probably one of the first essential steps toward generating and preserving life. The "insides" and the "outsides" of a cell must differ from each other lest they get into a state of *chemical equilibrium*, which as we previously saw, is a "sciencey" way to describe death (as in the earlier example of the house with no heat in the wintertime). Biological membranes are the cellular structures in charge of that separation. Some of the most important substances that we'll find on either side of the cell membrane include *ions*, which are electrically charged atoms. If an ion possesses more electrons than protons, its net charge is negative. In contrast, if the ion has more protons than electrons, the ion is positively charged. At the most fundamental level, bioelectricity has to do with how cell membranes keep the internal cell environment and everything outside of the cell separate and distinct, particularly with respect to ions and their associated electrical charges. The relative amount of negative or positive charges in either side of the cell membrane is sometimes termed the *electric potential* of the cell. In essence, bioelectricity is the production of such an electric potential by living organisms. These electric potentials control virtually all physiological activity in living beings, and some of these activities are truly unusual!

Just as the electric potential in an electrical plug must somehow be channeled to a particular machine to make it work, the electric potential in a living cell powers up life-related activities. Incidentally, when a cell displays an amount of charged particles that differs from the charges present in the external environment, we can say that such a cell is *polarized*.

The electrical charges present in polarized cells that eventually result in bioelectricity are tiny voltages for sure. The difference in charge, or in other words, the voltage difference between the inside and outside of any cell, is measured in millivolts (mV; one mV equals 1/1000 volts). To give you some perspective on this voltage, we all know that normal batteries, like the ones in a flashlight or a remote control, usually carry a charge of about 1.5 volts (1,500 mV). Also, please recall that the common electrical outlet in the United States is about 110 volts (110,000 mV). As you can see, these cell voltages are tiny! I started this chapter by narrating a fictional account of the first time that people experienced a thunderstorm. Well, the voltage associated with a typical thunderstorm is in the range (give or take) of 100 million volts!

As small as a cell's voltages are, they are far from insignificant, as they are absolutely essential to keep the cell alive. Moreover, quite a few types of cells have developed the ability to manipulate the electricity within them and use it to transmit information. This is how muscles move, how hearts beat, and how nerve cells allow us to feel and think. These are just a few examples of what cells can do by manipulating electricity. For obvious reasons, these activities must be extremely well organized, and when this organization does not happen, bad things occur, including disease. How are certain cells able to use electricity not just to keep living but also to correctly perform a wide range of activities? Those cells that are able to use their bioelectrical properties to accomplish such feats are collectively called *excitable cells*. What we mean by the word "excitable" is that they have the ability to change the relative electrical charges coming in and going out in a controlled way, and they can use this electrical change to induce quite a few types of cellular events. The electrical properties that are closely associated with life are so important that many organisms have evolved their capacity to detect minute amounts of electricity. Not surprisingly, a great number of these are—you guessed it—fish!

DETECTING ELECTRICITY

Let's recall the primary mission of living organisms, which includes the effort to avoid becoming food for others. One of the most interesting abilities of some organisms on this planet is the capacity to detect and/or control electrical currents for defensive or offensive purposes, as we saw when talking about strongly electric fish. An all-important aspect of such an ability that we have not yet talked about is that even though any organism by its very nature works via electricity, almost by necessity, detecting or emitting electricity are abilities that only arise in an aqueous environment, as opposed to on dry land. This is because air is a poor conductor of electricity. For an organism to be able to detect or emit an electrical field in air, the electrical charges would need to be, well, lightning-like in magnitude.[52] On the other hand, every body of water on this planet is full of impurities, like minerals, for example. The presence of these impurities is what makes water excellently conductive, whether it is found in rivers, ponds, lakes, or the ocean. Even the water out of your sink faucet contains enough trace minerals and other dissolved substances to allow for good electrical conductance. Given the ubiquitous nature of electricity in living organisms, it is not so surprising that many aquatic species take advantage of their environment's affinity for conducting electrical currents and use it for survival purposes. There are a wide variety of organisms able to perceive electrical fields (one of the most exotic being the platypus, which we will see in an upcoming chapter, although in a different context). For simplicity's sake, we'll discuss this ability exclusively from the perspective of a fish.

There are many species of weakly electrogenic fish that have refined the rare biological art of sensing a variety of electrical fields, including those produced by other living organisms.[53]

This is, after all, probably the most basic way of taking advantage of the existence of naturally occurring electrical fields. The obvious advantage being that every kind of organism naturally emits an electrical field. No organism can choose to turn it off or hide it. By virtue of these ever-present electrical fields, any living being advertises its presence for the whole world to see—at least, the part of the living world with the ability to "see" electricity. An organism capable of detecting electrical fields would have a distinct advantage over other organisms because it will sense the presence of prey or predator. Upon detection of food or foe, the fish can go into "hunt mode" or alternatively can swim away, putting as much distance as possible between itself and danger, and therefore living to see another day.

There are two basic modalities for electroreception, namely the *passive* and the *active* modalities. Passive electroreception is simply the capacity to detect electric fields in the environment. Such an ability, in addition to helping detect other critters in the immediate vicinity, helps a fish navigate its watery world, especially when swimming in murky waters or when cursed with poor visual capacities (and usually a combination of both). In contrast, active electroreception is a much more directed process. In this case, the organism generates a targeted electrical field with the explicit purpose of scanning its vicinity. When another living being (which, as we saw, always has an electrical field of its own) gets within range of such a fish's *electrogenic* field, the fish is able to sense the interaction between the two fields and act accordingly.* Moreover, fish with sensitive enough nervous systems and electroreceptive abilities can actually distinguish between, say, a rock or a plant, another fish, or even smaller critters like yummy invertebrates. In addition to searching for food and navigating their environment, the generation and detection of electrical fields by these fish allows them to communicate with each other, to establish territoriality, and to find mates.

In a nutshell, passive electroreception is just "listening" to whatever electrical signals are out there. On the other hand, active electroreception is correctly likened to echolocation—the system used by bats, dolphins, and other, stranger organisms to listen for the echo of an emitted acoustic signal.[54] But in order for an organism to use the strategy of active electroreception, said organism must be capable of producing electrical fields of a significant magnitude, well beyond that intrinsic to the electrical properties common to all organisms on Earth.

Strongly electric fish were easily understood, as the usefulness of strong electric shocks for defense or to capture prey was pretty evident. Weakly electric fish were another matter. Early scholars were puzzled by the apparent uselessness of the small amounts of current

---

* I. Cannot. Resist. The fish senses a *"disturbance in the force . . ."*

these fish produced. Even Charles Darwin wondered about the purpose of such small fields and, in fact, he talked about these fish in the "Devil's Advocate" section of his best-known book, *On the Origin of Species*, as an organism that he found hard to reconcile with his theory of natural selection. Alas for Darwin, once we learned about the electroreceptive nature of such fish—mystery solved!

These days, you can even find some of these weakly electric fish at pet stores. Some of the most common belong to the *Mormyridae* family and are better known as elephant fish.* As the name implies, their snout is closely reminiscent of an elephant's trunk, and they are actually quite cute! There are close to two hundred described species of elephant fish, all native to African rivers and lakes. Their size ranges from about an inch to almost five feet long.

How are these and other fish able to generate their electrical fields? All electrogenic fish possess some version of the electric organ we talked about earlier in the chapter, as well as specialized body parts that allow them to detect electrical fields. Curiously, scientists discovered the anatomical structures that endow fish with such powers before people knew much about electricity in general, and certainly before they knew about bioelectrical phenomena. These structures were discovered first in saltwater fish, mainly sharks, rays, and skates. In 1678, Italian physician and scientist Stefano Lorenzini described a mysterious network of small, interconnected cavities filled with a jellylike substance within the tissues of certain fish. This network connected to the external environment through distinct "pores," of which there are several types. It is important to clarify that Lorenzini did not discover this network. Depending on the source, the credit for this discovery has to go either to the Italian scholar Marcello Malpighi (widely known as the father of microscopic anatomy) or the Danish scientist Niels Stensen (best known as Nicolas Steno, an accomplished polymath in his own right) in 1663 and 1666,

---

* These are the same fish that were trained to send electrical signals on command (page 49).

respectively. Nonetheless, even though Lorenzini was not the first to observe these structures, he studied them in quite a detailed way, and they are therefore called "ampullae of Lorenzini" in his honor. However, try as he might, Lorenzini could not figure out what were they for; it was a mystery to him, and their nature and function remained a mystery to everyone for about three hundred years more. It was not until 1960 or so when scientists elucidated their electrical detection properties. Most scholars believe that part of the reason it took so long to discover the purpose of these organs is that virtually all fish species have them, or something very much like them, whether said fish are strongly electrogenic or not. Interestingly, in addition to detecting electricity, these electroreceptive organs are also quite sensitive to small changes in temperature, which undoubtedly adds to their usefulness for the organisms possessing them.[55]

Electroreception seems to be a prehistoric ability. You will recall that ampullae of Lorenzini were first found in rays, skates, and, rather famously, in virtually every shark species. You see, sharks and their friends have been here for at least four hundred million years. Moreover, you will also find these ampullae in a slightly younger, but still ancient, fish lineage (about 360 million years old): the coelacanth line. Coelacanths (*Latimeria chalumnae*) are fairly famous fish, notable for being a dramatic example of a "living fossil." These big fish, between six and seven feet long, were known in the fossil record but thought to have been extinct since the time of dinosaurs, roughly sixty-five million years ago; this is the key to their claim to fame. In the 1930s, to the surprise of more than one scientist, a pretty much alive coelacanth was captured in South African waters. However, the fish in question was very much dead by the time it was noticed at, of all places, a fish market. Over time, scientists found several other coelacanth specimens and even discovered whole populations of them. They live and thrive at various geographical locations in the waters around Indonesia and East and West Africa. In fact, the ones that live in Indonesian waters belong to a second species (*Latimeria menadoensis*).

The discovery of the coelacanth's electroreceptive capacities was an interesting curiosity, but truth be told, not completely unexpected because their stomping grounds are very deep waters (at least about 9,000 feet deep). Virtually no light ever reaches them at such depths, and therefore electroreception would be quite handy to find what to eat, to know what to avoid, and again, to determine what to mate with. Come to think of it, the phrase, "what to eat, what to avoid, and what to mate with" rather neatly summarizes what life—and especially life in its biological sense—is all about!

SEEKING IMMORTALITY

One of the most intriguing aspects of modern research trends in bioelectricity is the role that electricity has in the biology of development and regeneration. Although frequently confused, development and regeneration are two separate, albeit quite closely related, aspects of biology, especially at the molecular level. A discussion of the particular similarities and differences between the two processes is beyond the scope of this book, but there are excellent general articles that you can check out if interested.[56]

A way of defining "development" is to state that it means "genes in action," and that is not a bad definition. But as important as genes are, they do not—they cannot—work alone. Genes isolated from the environment do not have any context to work on. There is a significant and still growing body of evidence indicating that an important part of the environmental conditions that directly affect gene expression include (this will not surprise you) electrical signals, which work closely with the genetic instructions of an organism to do all the things that life does.

A biological phenomenon closely related to development is the mysterious process of regeneration. Virtually every multicellular organism on the planet has at least some capacity to regenerate damaged tissue or body parts. For example, every time an organism heals in response to injury—as when you are healing from a paper cut (hopefully not one from this book!)—it is undertaking a kind of regeneration. Even some organs in the human body, such as the liver, display significant regenerative capacities. However, in general, humans have rather limited abilities in this area. A good rule of thumb is that the closer we get to the vertebrate line, the lesser the creatures' regenerative powers seem to be, with a few distinguished exceptions like some salamanders and other amphibians, as well as fish species like the zebrafish, which can regrow whole limbs, their eyes, and, in a pinch, even parts of their very hearts and brains. Without a doubt, the capacity for regeneration is one of the most exquisite survival strategies ever evolved on this planet, and there are some curious organisms out there that are true champions as far as regeneration is concerned. These include several species of sponges that possess extreme regenerative powers. For example, if you take one such sponge and completely dissociate its cells and then—to add a literal insult to an actual injury—let's say, pass those cells through a mesh, those cells, if left alone, will eventually reunite and reform a complete sponge, a normal, healthy sponge, none the worse for wear. True, sponges, as interesting as they are, are barely animals at all (with apologies to all my zoologist friends), as they do not seem to do much (I am now talking about the sponges, not about my zoologist friends).

A more complex organism capable of achieving the same feat— cell dissociation, mesh, the works, followed by complete reconstitution of the animal—is a freshwater invertebrate called *hydra*.[57] These are small Medusa-like organisms—most are only a few millimeters long—whose name comes from the ancient Greek myth. The actual hydra belongs to a group of animals called cnidarians, which includes, among others, jellyfish, sea anemones, and corals, all of which are capable of some degree of regeneration. Keep in mind that the mythical hydra was no jellyfish! It was a giant serpent-like monster with multiple heads. In the myths, if a hero (*hero*: the guy who survived to tell the tale) or a regular guy (*regular guy*: the guy who did not survive, just like a movie extra, or the famous—but also mythical—red shirt of *Star Trek* lore) managed to cut off one of the heads, the hydra was able to grow a new one in its place. Sometimes, it even grew more than one head to replace the head it lost. The actual hydra, although thankfully not a giant monster, is a real-life predator and a fascinating organism in its own right, and it truly can regrow lost body parts.[58] And in contrast to sponges, hydrae are very active animals. They move about and actively capture prey. Not surprisingly, they have a relatively complex nervous system, which can fully regenerate in the reconstituted organism. Now this is getting truly interesting from our human point of view, because functional nervous system regeneration is one of several "Holy Grails" of the biomedical sciences. Think about the many, many people who suffer from neurological diseases or who have damaged their brains or other parts of their nervous system as the result of an accident or a stroke. In this sense, hydra studies may prove quite useful: if we ever hope to help patients with injuries like these, we need to learn how to regenerate working nerves and nervous tissue. We could prevent so much suffering! However, there is a caveat: the hydra has a nervous system . . . but it does not have an actual brain. This means that whatever we learn from nerve regeneration in hydrae may well not apply to organisms with a brain. What do we do then?

Do not fear, my intrepid reader—there is an animal in existence that combines rather powerful regenerative abilities with the possession of an actual brain. They are probably my favorite nonhuman organism, and I briefly mentioned them in the previous chapter. They are called planarians, a type of flatworm that is widely distributed over this planet. There are freshwater, marine, and terrestrial species, all generally small in size, many in the centimeter range, though some species are much smaller and some are much bigger.

I do know that worms are not everybody's cup of tea—for the record, I love them!—but what makes many of them so interesting, among other things, is their often impressive capacity for regeneration.[59] However, they cannot—as hydra and some sponges can—withstand cells dissociation, etc., as we described before. If we were to put planarians through that, we'd have nothing but worm soup.

Nevertheless, if you cut a piece of a planarian as small as about 0.03 cubic millimeters[60] (which is smaller than a grain of salt), over time that tiny fragment is able to regenerate into a whole organism. I have said elsewhere that this is much like a human losing a hand, and that hand in turn growing back into a whole human, new brain included.* Well, planarians can do this, brains and all! And not just any brains, mind you, but brains with the capacity to generate relatively complex behaviors (for a worm that is).

Hydrae, planarians, etc., are able to achieve full regeneration not under laboratory conditions, but rather *in their natural environment*, meaning that if a predator attacks a planarian, for example, and leaves a fragment as small as the ones we talked about, that fragment, if left alone, eventually will end up as a complete animal. This qualifies them not just as masters of regeneration:[61] in no small part courtesy of bioelectricity, they are masters of survival. Even though the precise way in which electricity affects regenerative processes is not entirely clear, the best current evidence indicates that electricity is one of the main players that drives these processes. I am sure that future research will bring many surprises! There is an additional aspect of planarians, hydra, and related animals that makes them even more unusual. One of the most amazing characteristics of some species of hydrae and planarians is that they do not seem to go through senescence (in other words, they do not seem to age). The simple fact is that nobody knows for sure how long these guys naturally live. It stands to reason that all the longest-living animals tend to be the simplest ones, but it is one thing to live for a long time and quite another to escape old age completely.

A particularly interesting and extreme example of a living being that seems to escape aging is a rather small but pretty type of jellyfish called *Turritopsis nutricula*. Many of us, when going through difficult times in our lives, wish for the simpler, carefree days of childhood, am

---

* Time Lords seem to be able to do just that. If you do not know what I am talking about, go watch *Doctor Who* at once! I'll wait.

I right? Well, *T. nutricula* is capable of doing something very much like going back to infancy. This species of jellyfish hatches normally, lives normally, grows normally, undergoes metamorphosis, achieves sexual maturation, and eventually reproduces. As adults, they settle down and form colonies that further reproduce asexually by budding,* nothing unusual in that. However, if an individual *T. nutricula* is exposed to stressful stimuli, the jellyfish is able to revert back to its younger self. Mind you, it does not become a smaller, or even younger jellyfish, but an actual jellyfish *embryo*. Once the environmental stress is gone, the embryo starts developing and growing normally again, therefore being able to give itself a second chance at life (and potentially a third, a fourth, etc.; we do not currently know if there is a limit to their ability).[62] How long do you think it will be before somebody finds that bioelectricity has something to do with this jellyfish's ability to escape old age? My unapologetically educated guess is: *not long*. As you might imagine, planarians, hydrae, and now these immortal jellyfish, among other organisms, are under intense investigation to discover how they resist the passage of time. Of course, this is something that people have aspired to achieve ever since we became conscious of our inevitable mortality.** The tale of this jellyfish is a developing story, as it were, and it is bound to be one worth watching!

ELECTRICITY AND LIFE

As we briefly explored in the previous section, there is a rapidly growing body of evidence that indicates that the bioelectrical properties of cells and tissues play an essential role in development, regeneration, and tissue repair. In brief, it seems that by controlling the bioelectrical

---

* This is a form of asexual reproduction where an organism "sprouts" offspring from its body.
** Hence the multiple legends across time and cultures telling of fountains of youth, of wishes granted by supernatural beings, and many other variations on the theme.

properties of cells, we can manipulate and even sometimes override the genetic expression of an organism (in other words, its phenotype) without changing its genetic information (its genotype) at all.[63] At the cellular level, one of the main players controlling the flow of electrical currents between cells is a structure that is technically called a "gap junction," which we can compare to a small tunnel. It seems that by precisely controlling the electrical current between cells, one can control how an organism develops and regenerates, as in the following striking example.

There is a recent study using (what else?) planarians, in which the researchers cut the miraculous little buggers and allowed them to regenerate as usual. As expected, if left alone, they regenerated normally according to their particular genetic instructions, which are specific to each unique planarian species. Straightforward enough, right? Well, they repeated the experiment, but this time exposed the regenerating planarians to a drug that blocks the aforementioned gap junctions, thereby disrupting various cellular electrical processes. In this experiment, most planarians regenerated, all right, but something quite curious happened. A significant fraction of the worm population regenerated heads that looked like those of *other species* of planarians. It gets even stranger; the heads of those "wrong" planarians returned to their original shape in a few days, once the scientists withdrew the gap junction drug from the planarian's environment.[64]

Please allow me to offer some perspective on these results. Let's suppose that we try these experiments with mice, for example (rodents cannot regenerate their heads; this is a thought experiment—for humans). In this imaginary experiment, some of the mice would regenerate normal mice-like heads, while some of the mice would regenerate squirrel-like heads, some others would regenerate ratlike heads, and yet some others would regenerate chipmunk-like heads *without changing their genomes whatsoever*, in other words, only mice genes, no rat-, squirrel-, or chipmunk-specific genes at all. Furthermore, as in the actual planarian experiment, if we take away the drug that affects gap junctions, the "wrong-headed" individuals would redevelop mouse-like heads, as if their genome were reasserting itself. Well, how could this be? The fact is that in addition to the genetic code, there seems to be also a bioelectric code[65] that works together with the genome and the environment to determine an organism's anatomy and physiology according to the electrical environment within cells. Isn't this *wild*?*

Just imagine the future implications and possible applications of such knowledge. Think about all the people who die waiting for an organ transplant, people who have brain damage for a variety of reasons, and all the people with currently intractable genetic diseases. The more we learn about what controls development and regeneration—and how—the closer we get to helping such patients.

Biology never ceases to amaze me, and if I am doing my job right, it should begin to amaze you as well, if you weren't astounded by it already.

---

* I narrate this interesting story in more detail in my recent article: Pagán (2017, *in press*), "Planaria: an animal model that integrates development, regeneration and pharmacology." *The International Journal of Developmental Biology*, doi: 10.1387/ijdb.160328op.

CHAPTER 4

UNUSUAL SUSPECTS

We are all very familiar with a wide variety of toxic and venomous animals—the "usual suspects," if you will. The organisms that usually come to mind when thinking the word "venomous" are snakes, spiders, scorpions, and the like. Other animals that by definition are venom-producers include bees and wasps, centipedes, certain types of ants, and various similar critters. As remarkable as all of these organisms are, nature has some still stranger specimens in store for our learning pleasure. Let's meet some of these "unusual suspects!"

» VENOMOUS DINOSAURS?

Very few things are scarier than a big, angry lizard trying to bite you, with the probable exception of a big, angry, and *venomous* lizard trying to bite you! We do not usually associate lizards with venoms, like we do snakes. Venomous snakes are rather common; in fact, they are so common that most people's first reaction when encountering an unknown snake is to worry about whether it is venomous or not. In contrast, until very recently, only two types of lizards were actually found to be venomous, both of them belonging to the *helodermatid* family. These are the Mexican beaded lizard (*Heloderma horridum*), and its close relative, the much better known Gila monster (*Heloderma suspectum*).[66] The Gila monster and its relatives grow to about two feet in length (yes, they are big lizards). However, they are rather slow, sluggish even, and if you stay out of their way, they'll stay out of yours.

The term *Heloderma*[67] in these lizards' scientific name refers to the appearance of their skin, which is covered with scales that look like "nail studs" or "rivets."[68] As a matter of fact, their skin is rather beautiful, patterned in alternate black, yellow, and orange scales. These lizards are widely distributed in the desert environments of the southwestern United States and Mexico as well as parts of Central America, found as far down as Guatemala. The two species diverged from a common ancestor about five million years ago, making them closer relatives than us humans and our closest cousins, the chimpanzees. (Humans and chimpanzees diverged from a common ancestor about seven million years ago.) Interestingly, *helodermatid* lizards are more closely related to snakes than to other lizard species, although their closest lizard relatives are the varanids. (Varanids include the well-known Komodo dragon and other sister species. *Helodermatids* diverged from varanids some fifty million years ago.)

The first written historical account of helodermatid lizards dates from 1577, but Gila monsters were part of the folklore of many indigenous cultures way before then. As with many animals, the relationship between humans and Gila monsters is a complex one, painted very vividly by the human imagination. Many cultures attributed mythical characteristics to these lizards. Traditionally, some Native American tribes asserted that even the mere breath of the Gila monster was lethal. What's more, Gila monsters have had an undeserved reputation for violence. In reality, these animals are not aggressive at all, and virtually every well-documented account of a Gila monster's bite is due to the human mistreating or provoking the lizard. That being said, when they bite, they *bite*! They hold on to the attacker with a tenacity seen in few other animals; they just don't let go.*

Despite all this, not all legends related to the Gila monster are negative. For example, certain legends of the Tohono O'odham Native Americans tell the story of how these lizards developed their distinctive skin patterns. One story tells of the first saguaro wine festival,[69] which legend says took place hundreds of years ago. In addition to the people who attended the festival, all the animals were invited. As expected, everyone, human or nonhuman, wanted to look their best for the occasion and the Gila monster was no exception. It finally hit upon an idea spectacular enough and proceeded to collect brightly colored pebbles from the desert. With those pebbles it made itself a beautiful—and at the same time, hardy—coat, which it wore to the event and has worn ever since because it liked it so much.

In quite a few indigenous cultures, people also attributed medicinal properties to these lizards, anticipating their role in modern medicine. The Navajo tradition considers the Gila monster as the first shaman, or medicine man.[70] This is interesting, because this lizard is the source of a compound that is now an established part of medical practice, often used to treat a subset of diabetic patients who tend not to respond to the more traditional therapeutic options currently available.[71]

Recent scientific work has provided information indicating that there are a surprising number of other venomous lizards as well. You probably have heard of one of these, which I mentioned briefly previously, the aptly named Komodo dragon, a native of Indonesia. Yes, *that* Komodo dragon, the one that can be about seven feet long or more. The venomous nature of Komodo dragons was discovered only recently.[72] Prior to these discoveries, we knew that a Komodo dragon bite was a serious thing, and likely fatal too, but most assumed that the nastiness of its bite had to do with the many disease-causing

---

* In fact, the worldwide expert on these lizards, Dr. Daniel D. Beck, has nicknamed them the "pit bulls of the lizard world."

bacteria that these lizards host in their saliva. Now it is clear that Komodo dragons and some smaller versions of the species do have actual venom glands that produce proper venom with a variety of biochemical properties.* Chillingly, recent research seems to indicate that a much bigger relative of the Komodo dragon, a large prehistoric varanid lizard named *Megalania* that was about twenty feet long, was also venomous.[73] Yikes! Imagine a venomous, crocodile-sized beast. Speaking as a biologist, I can't believe that I am saying this, but here it goes: I am kind of glad that *Megalania* is extinct.

As I was researching this book, it surprised me to learn that most lizard families have representatives possessing venom glands. In fact, venomous lizards predate venomous snakes by millions of years and have a distinguished biological history, which, depending on the source, seems to date from the time of the dinosaurs. One of these examples is the tuatara, an ancient type of lizard that hails from New Zealand and has been knocking around for the last two hundred million years or so. By the way, speaking of dinosaurs, for some reason, 2010 was a big year as far as venomous dinosaurs were concerned. In that year, two publications reported possible venomous species. In the first case, the scientists examined the nearly complete skull of an avian (birdlike) dinosaur called *Sinornithosaurus,* which was about 145 million years old. Now, before all of you dinosaur enthusiasts delight in the thought of something like a saber-toothed, venomous *T. rex, Sinornithosaurus* was not a big animal. It was, in fact, about the size of a chicken** and it was an important fossil even before people thought it was venomous because it was one of the first organisms that developed feather-like structures. At any rate, the researchers inferred that this animal was venomous based on three main observations: first,

---

* And that the types of bacterial species in their mouths are not significantly different from the bacterial population in the mouths of other reptiles.

** I cannot stop thinking about an angry, poisonous chicken. Yes, I have the humor of a six-year-old.

it was a *heterodont*, (this means that some teeth are significantly longer than the others); second, some of these longer teeth had grooves reminiscent of the ones found in the teeth of living venomous lizards; and finally, the skull itself had some structural features that were interpreted as possible locations of venom glands.

At the time of this report, these were exciting findings. Who does not like dinosaurs, let alone venomous ones? However, one of the best characteristics of science is that it is self-correcting. A few short months after the publication, another group disputed the original interpretation, meaning that they thought that *Sinornithosaurus* was not venomous at all. To date, the question of whether there were venomous dinosaurs seems to be an open one. Or is it? In another 2010 report, scientists documented the finding of the fossilized teeth of a dinosaur even older than *Sinornithosaurus*. This one was called *Uatchitodon*, dated to about two hundred million years old. Its teeth showed the characteristic grooves associated with the teeth of venomous reptiles, but even better, some specimens displayed complete tubelike structures within its teeth, structures reminiscent of modern snake fangs. Alas, the only remains that we have of *Uatchitodon* are its teeth; no one has discovered any other body parts of the reptile. It would be interesting to examine a complete skull to try to determine if *Uatchitodon* also had venom glands—so far, it seems to be the best available specimen that offers evidence for the possibility of venom in really ancient reptiles. One thing is for sure: reptilian venom is not a recent invention of nature.

POISONS, TOXINS, VENOMS

Whenever we think about biological defenses, the mind goes naturally to the topic of venoms, toxins, and poisons. No self-respecting book on the survival of organisms is complete without mentioning this trio. Confusingly, all too often we use these three terms interchangeably, but they are in fact subtly different.

Poisons: A poison is any chemical substance or mixture of substances that negatively affects the physiology of a living organism. It is essential to note that practically anything can act as a poison if ingested in large enough quantities—to paraphrase Paracelsus (1493–1541), the well-known father of toxicology. Just to give you an example, water itself, despite being considered pretty much the least toxic substance known, has been the cause of several deaths due to overconsumption.[74]

Poisonous materials can include heavy metals like mercury, arsenic, and cadmium; gases like carbon monoxide; and natural and synthetic chemical carcinogens of various types, among many other substances. Although the body can tolerate these compounds in reasonably small amounts, higher amounts can be quite toxic indeed. However, despite their evident toxicity, we do not call these substances "toxins." Rather, we reserve the term *toxin* for poisonous substances *made by or derived from biological sources*, that is, substances made as part of the biochemistry of a living organism.[*] From now on, we will only talk about toxins within the context of the living world. This is why, later, we will describe molecular oxygen as the first toxin, as the relatively high amount in our planet is directly attributable to the chemistry of life itself. That being said, the early history of our planet is full of poisonous substances, and even though molecular oxygen seems to have been the first toxin made by life, it did not remain the only such substance for long.

Toxins: Biological toxins include some of the most interesting compounds in nature. Although each is a product of an organism's metabolism, some are structurally complex while others are quite simple. In either case, their synthesis is usually subject to intense

---

[*] Incidentally, the word "toxin" itself has an interesting origin. It comes from the term "toxicon," which in turn means "arrow poison." This came to us from—who else?—the ancient Greeks.

evolutionary pressures. This makes some cells in quite a few types of organisms true master pharmacologists, with an exquisite series of skills that are beyond the abilities of any human organic chemist.

Consistent with the overall theme of this book, toxins make up some of the most effective mechanisms of defense, competition, and predation to be found in nature. Many species that lack evident defensive characteristics like sheer size, brute strength, fast speed, the ability to fly, or even the capacity of intelligence, have nevertheless learned to use biological toxins to their advantage. Depending on the specific context, we'll call them poisonous, toxic, or venomous organisms.

Virtually every type of organism on our planet has at least one representative capable of producing toxic substances. In the words of the famous venom researcher Dr. Bryan Fry, *"where there's life, there's venom."*[75] I tend to agree; this strategy is one of the most widespread tactics of biological defense. More specifically, formally defined, the concept of biological defense through chemicals occurs when *"an individual contains or uses behaviorally one or more chemical substances that deter predators and/or parasites."*[76]

In general, a poisonous or toxic organism stores toxins throughout its body, although in some cases such toxins concentrate at specific locations in the plant or animal. For example, a plant may have toxic seeds but edible fruits and vice versa. Other examples of this include the accumulation of toxins in specific organs of the fish used to make fugu, as discussed in Chapter 1.

> *"If you bite it and you die, it is poisonous; if it bites you and you die, it is venomous."**

The contrasting strategies employed by organisms that use simpler toxins as opposed to creatures that utilize venoms is most likely the origin of this wry phrase. One of the defining characteristics of

---

* Or alternatively: "Single toxins are ingested, venoms are injected."

toxic organisms is that they have no specialized delivery mechanisms. Their defensive strategy is purely passive—the toxic creature requires the "cooperation" of the organism that preys on it. This usually entails the predator trying to take a bite of the toxic organism. That is why, for example, we usually hear about "toxic" or "poisonous" plants as opposed to "venomous plants," since generally a plant does not deliver its toxin to an attacker in an active fashion. Rather, the predator goes to the plant with the worst of intentions (to eat it), and it is only then that they get their toxic comeuppance.

Toxicity is a great representative example of how, so often, "nature invented it first." Take insecticides, for instance. Our agricultural sciences use (and oftentimes overuse) many chemicals to control pests. Humans did not invent this practice. Many plant predators are insects; unsurprisingly, quite a few plant-derived toxins are considered the first examples of chemical pesticides. You may be surprised, however, to learn that these naturally occurring insecticides include several substances that are abused by humans, such as the alkaloids cocaine, morphine, and nicotine, among many others.[77]

Venoms: Venoms differ from toxins in several key aspects. A really nice definition for the term *venom* is a "*secretion, produced in a specialized gland in one animal and delivered to a target animal through the infliction of a wound, which contains molecules that disrupt normal physiological or biochemical processes so as to facilitating* [sic] *feeding or defence by the producing animal.*"[78]

Venoms invariably consist of complex mixtures of many, in some cases hundreds of, different compounds. Some of these compounds are toxins in their own right while other substances, despite not being toxic themselves, are nonetheless able to fine-tune the toxic response. Incidentally, toxic/poisonous/venomous organisms generally become so by two routes: *genetics* or the *environment*. As in biological life in general, the apparent dichotomy between nature and nurture here is at best a misleading one. The genetics part is straightforward. Many organisms possess genes encoding enzymes that produce toxins; sometimes there are genes that encode the toxins themselves. Alternatively, organisms can obtain toxins environmentally by consuming other (toxic) organisms or by keeping a symbiotic relationship with properly toxic beings. This begs the question: How is it that certain critters tolerate the presence of a toxin in their body? Or, how do organisms become resistant to their own toxins? The answer is usually biochemical, namely mutations that modify crucial regions of the toxin's target at the molecular level. There are many documented examples of such mutations, and these are essential components of the evolutionary arms races that we discussed in Chapter 1.

As mentioned, all venom-producing organisms have special anatomical features that usually include specialized organs that contain, store, and deliver the toxic mix. Such a mix is oftentimes called a "toxic cocktail" on account of its chemically complex nature. We usually call the organs that store the venom "venom glands." These glands work closely with muscles that help "pump" the venom to a complementary set of structures that include stingers, grooved teeth, or even hollow fangs, which are in charge of the actual delivery of the venom. When working together, the venom gland and the stinger or tooth actively deliver or even inject the venom into the intended target in a highly efficient way.

NATURE'S NEEDLES

These venom delivery mechanisms represent yet another invention of nature that predates the human version of it—namely, the hypodermic needle—by millions of years. To give you an idea of how recent are human-made hollow needles as opposed to nature-made ones, the history of medicine records that the modern hypodermic needle is not quite four hundred years old, and it was developed over a period of about two hundred years. In 1657, the famous British architect and polymath Christopher Wren was the first to administer an intravenous injection of a compound (actually, an anesthetic) to another living being—a dog. The canine came through it with no apparent ill effects, so Wren decided to try the procedure on a person. The subject (a man, by the way) obligingly fainted as soon as the needle went in—no anesthetic necessary!

The strength and potency of venoms depends as much on the amount produced as on their chemical complexity. In general, however, the more complex the venom, the more effective it is. What do we mean by "venom complexity"? Scientists define such complexity in terms of how many different active chemicals may be present in a given venom mixture. The main idea is that when a predator injects venom into a prey, each of the toxic components of the venom affects a (more or less) specific physiological process of the target organism. Therefore, a complex venom blend will potentially interfere with multiple aspects of the prey's physiology, not just a single one, thereby maximizing the chances of a successful attack—and making successful defense against it that much more complicated as well. Undoubtedly, the chemical variety present in venoms is the main reason that venoms are such a rich source of interesting and even useful compounds, including some that have already found a place in our medical sciences and many others that are quite promising in this respect. There are quite a few examples of medications whose "origin story" is traced back to venoms and toxins. For example, the whole field of local anesthesia unarguably began with a plant toxin: cocaine.[79] Some commonly used muscle relaxants were inspired by the effect of curare, another plant toxin, on muscle contraction.[80] Animal toxins have also contributed to medicine. A particular type of toxin from cone snail venom (we'll see these snails soon) is being used "as is" to treat patients with otherwise intractable pain.[81] And finally, even the aforementioned Gila monster has a part in this story, as one of its venom components is a toxin that inspired an antidiabetic medication.[82] There are many other examples as well!

» POISONOUS* FROGS

You may have heard of poison arrow frogs. These are generally small critters from South and Central America, which are brightly colored and beautifully patterned. This coloration seems to be used by the frogs to advertise their toxicity to potential predators. Why else would they "want"** to stand out in their environment? After all, they are not big, fast, or fierce. Using the language of nature, essentially they are saying: "*Eat me at your own risk; you'll pay for it soon enough, you'll see.*" Usually, when organisms display colorful patterns in order to advertise their toxicity or unpalatability, we say that such organisms display *aposematic coloration*. Interestingly, the same coloration patterns can serve to attract mates of the same species. Isn't it remarkable that exactly the same signal serves to repel or attract, depending on the context? Most likely it is an evolutionary gamble. The organism risks predation for the possible increase in its reproductive success. This is a common strategy used by many life forms, such as peacocks, for example. If you see pictures of these tiny arrow frogs, you will see that they are especially beautiful. In addition to their bright coloration, many species display various combinations of striped or spotted patterns. Their beauty hides especially nasty toxins that are usually

---

* They are technically toxic, not poisonous, but when a name sticks, it gets stuck.

** Please remember that these organisms do not "decide" to do anything in the cognitive sense. We have to be careful about personifying them.

located on the frog's skin. This dangerous beauty reminds me of a quote by Dr. Edward O. Wilson:

> *"If a small and otherwise unknown organism is strikingly beautiful, it is probably poisonous; if it is not only beautiful, but also easy to catch, it is probably deadly."*[83]

Poison frogs are frequently called the "most poisonous animals alive," and for good reason too.[84] Just how poisonous are these little frogs? Well, a typical poison frog has enough toxin to kill about ten grown men. Assuming that a typical man weighs close to 200 pounds (about 90 kg), this translates to about 257 typical laboratory rats (about 350 grams per rat) or to about 2,250 typical laboratory mice (about 40 grams per mouse). These little guys pack a punch!

In many cases, scientists have established that the frogs themselves do not make the toxins. Rather, they seem to acquire them from the environment, as evidenced by the fact that in general, when in captivity, the frogs are not poisonous. Their toxins have been found in various types of insects that live in their natural environment, and these toxins display a great deal of structural and physiological variability. I am always simultaneously amused and concerned when I see nature documentaries or pictures showing this type of frog held by people in a nonchalant, barehanded way. As I said above, they store their toxins in their skin. If you hold one of these frogs and by chance have an open cut or abrasion of your own skin, the toxin will likely get in you, with less than pleasant consequences. When native South Americans coat their arrow tips or darts with the frog's poison, they do not even touch the frog directly, as they know better. Instead they use a short wooden stick to hold the critter in place while rubbing the dart or arrow tip against the frog's skin. Upon this stimulation, the frog becomes agitated and releases a skin secretion laced with toxins. The hunters cover their darts with the secretion and release the froggies once they're done with them, usually after two or three "milkings."

Ever-creative nature offers more examples of unusual toxic/venomous amphibians. You already read about a specific example of a quite poisonous newt within the context of evolutionary arms races, but now I want to talk about another newt species, the Spanish (sometimes termed Iberian) ribbed newt, *Pleurodeles waltl* (briefly mentioned in the earlier explanation of scientific names). As the common name implies, this newt is found mostly in Europe, especially in Spain and Portugal, although its distribution extends to Morocco and nearby places. A related subspecies is thought to live deeper in the African continent, but actual data is scarce at the moment. *P. waltl* is a medium-size newt, with an average length of about eight inches.[85] It is also toxic, but as we saw before, this is not as uncommon as you would think. What makes this species particularly curious is that when threatened, it is able to make some of its ribs pierce its own skin, forming a kind of spiked armor, which—as it comes out—gets coated with toxic secretions from the newt's skin. This is reminiscent of the practice of coating spears with poisonous coral (or rubbing them against the skin of a frog, for that matter). As expected, this mechanism used by Iberian newts is very effective at deterring predators, and ingenious too; their ribs have fully formed and functional joints that allow them to rotate and therefore protrude through the newt's skin at the appropriate angle. This behavior was described in the early 1800s, but it was only carefully studied in 2010.[86] In addition to being unusual, this is of considerable practical interest because nobody really knows how the newt avoids its own poison or how it is capable of avoiding infections every time it breaks its own skin. This newt does not appear to be unique, however: two related species (*Echinotriton andersoni*—the very aptly named "crocodile newt"—and *E. chinhaiensis*, both predominantly found in Asia) seem to use a similar defensive mechanism.

Another fascinating example of toxin use in nature comes courtesy of yet another two amphibians. There are two species of tropical frogs (*Corythomantis greeningi* and *Aparasphenodon brunoi*)[87] that make use of a strategy not unlike the example above. These frogs have fixed bony spines that protrude from their skull. Moreover, these are associated with proper venom glands, which makes these frogs not merely toxic—rather, these structures earn these amphibians the title "venomous." It seems that these are the first examples of *venomous* frogs found in nature, but more research is underway. It gets better (or worse, if you are on the wrong side of the spines). Preliminary studies of their venom indicate that it is several times as toxic as the pit viper venom, which is exceedingly dangerous in its own right— please remember that rattlesnakes are pit vipers. At least four other species of frogs are described as having head spines, but their venom has not been fully studied at the time of this writing.

So far we have seen a few examples of vertebrates that display toxic or venomous properties. As interesting as they are, these are not totally unexpected. Ah, but nature is not constrained to merely what we can conceive or imagine! There are indeed more organisms out there that have evolved venoms and toxins. As we'll see later on, some of these organisms even include birds and mammals! That's right, you do not have to be a reptile or an amphibian or have six, eight, or even a hundred legs to be venomous or toxic. Furry and cute little things (or just furry, not necessarily cute) can be venomous or toxic as well.

THE TRUTH ABOUT OXYGEN

The best science available seems to indicate that our universe is roughly fourteen billion years old, and Earth has been around for about 4.6 billion of these years. Life itself is not much younger than the age of our own planet—assuming, of course, that life originated right here on Earth. The best current evidence tells us that life appeared here as soon as it could. The event—or series of events—that originated life on our planet likely happened after the Earth cooled down enough from the intense and rather violent series of incidents that led to planetary formation in our solar system. This cooling down took place about 3.8 billion years ago.

As you may imagine, the environmental conditions of our planet just before life appeared were much different than they are now. Take, for example, the composition of the atmosphere. We think that the likeliest composition of the atmosphere then was a mixture of carbon dioxide, molecular nitrogen, and trace amounts of other gases. I want to bring your attention to a very important point: What gas is missing from this picture?

If you think about the current chemical profile of the air around us, the answer will jump right out at you: the missing component is none other than good old molecular oxygen (O_2)! This gas is currently essential for the survival of most types of life forms on our planet, particularly the multicellular ones, including us. Nowadays, our atmosphere is composed of roughly 21 percent oxygen and 78 percent nitrogen. The remaining 1 percent of the atmosphere contains water vapor, carbon dioxide, argon, and a variety of other trace gases. The fact that there is so much gaseous molecular oxygen here and that this relatively high amount is stable over time is quite remarkable. There are two main reasons why this is so unusual. The first reason is that in contrast to gaseous nitrogen (N_2), molecular oxygen is a rather active molecule, meaning that it likes to react with many other chemicals in the environment. We can easily visualize oxygen's molecular reactivity with a phenomenon very familiar to us. If you take an apple, cut a piece of it, and leave it on an air-exposed surface, in a very short time its pulp will "brown." In other words, the oxygen in the air combines with organic molecules in the fruit's pulp, *oxidizing* these molecules. This is essentially the same phenomenon as when metal rusts. Thus, the natural consequence of oxidation illustrates the tendency of oxygen to combine itself with other compounds. In the process, the amount of oxygen is gradually depleted until it is completely used up. Since oxidation tends to use up all free oxygen available, and oxygen is very reactive, to keep its levels stable at about 21 percent over geological time periods obviously requires some kind of mechanism to continuously replenish it. This is the second unusual aspect that accounts for the continued presence of molecular oxygen in our atmosphere: There is no known purely chemical way to sustain this oxygen concentration for millions, let alone billions, of years. This is a job for biological life!

Most of the free oxygen around us originated from living organisms through an extraordinary process called *photosynthesis*. Simply stated, photosynthesis is the use of radiant energy (light) of various wavelengths, depending on the exact nature of the organism, to break down a water molecule into hydrogen and oxygen. Hydrogen goes to power up certain reactions that produce chemical energy, and in turn that chemical energy is channeled to build complex organic (carbon-containing) compounds with carbon dioxide as the main ingredient. But, where does oxygen come in? Well, oxygen is merely a waste product of photosynthesis! It is as simple as that. The unadorned truth is that from the point of view of certain early forms of life, our dear oxygen was good for literally nothing. And in fact, oxygen was toxic to the life present in those times, and still is to some modern organisms. Molecular oxygen is a quite poisonous chemical species, you know. The very reactivity that makes oxygen brown an apple or rust some iron can wreak havoc on an organism's physiology. In fact, many scientists describe the series of events that made free oxygen so abundant in our planet as the "Oxygen Holocaust."[88] In the context of the history of life, this event led to the extinction of many species of microorganisms (as there were only microbes in the early days of life on Earth). However, in a paradoxical way, the increase of oxygen levels in our atmosphere eventually came with an unexpected benefit, the creation of the ozone layer about two billion years ago. Ozone (O_3) is an alternate form of molecular oxygen that is mainly present in the upper atmosphere and blocks a significant portion of the sun's ultraviolet radiation. This blockage is essential for the survival of many life forms, including us.

The above account of the history of how our oxygen came to be is, by necessity, a quite abbreviated story. A much more complete history of oxygen and its relationship with life is told elsewhere.[89] That being said, the take-home—and very surprising—message of this section is that without a doubt, the very same oxygen that is so essential to most life on Earth now was probably the very first biologically produced toxin. Indeed, biology is full of surprises.

» VENOMOUS CRUSTACEANS

Most multicellular organisms on our planet are *arthropods*, which are basically organisms possessing an exoskeleton and jointed limbs. Arthropods include a wide variety of living beings; some estimates indicate that close to 80 percent of all animal species on Earth belong to this category of organisms. The arthropods can be broadly divided into four main types: *insects* (ants, bees, beetles, etc.), *arachnids* (spiders, scorpions, and ticks), *myriapods* (centipedes, millipedes, and similar organisms), and *crustaceans* (crayfish, crabs, lobsters, and shrimp, among others). As you can imagine, many arthropods are of interest because of their agricultural importance. However, more significant from our perspective is the fact that quite a few species (and this is an understatement) have the ability to produce venom or are toxic in some way or another. Of the four general types of arthropods mentioned above, only the crustaceans did not seem, until very recently, to have a venomous representative. There were quite a few examples of *toxic* crustaceans, which are known to store various toxins derived from environmental sources. However, we did not know of any *venomous* crustacean, capable of actively injecting venom as a predatory or defensive strategy. Now we have at least one! This is one of the many reasons why I love science, especially biology; virtually every day you wake up and learn of a curious or exciting—and often-times both—discovery.

The only example of a venomous crustacean so far is an organism belonging to the rare class of the remipedes. These are small (typically less than two inches long) centipede-like marine animals that were previously known only as fossils until 1981, when live specimens were collected in an underwater cave in the Bahamas by Dr. Jill Yager,[90] who at the time was at the Department of Biological Sciences of the Florida Institute of Technology. Remipedes were not very well-known then (and perhaps even now), in part because they tend to live in underwater caves, which constitute a somewhat underexplored habitat. At the time of the initial discovery, the animals were unambiguously recognized as remipedes, but the specific species was not known. Flash forward to 2014, when a particular species of remipede, *Xibalbanus tulumensis* (*Speleonectes tulumensis* according to some sources),[91] was described as the very first venomous crustacean. The scientist who led the team that made this discovery was Dr. Björn M. von Reumont, from the Department of Life Sciences of the Natural History Museum, London. Dr. von Reumont's team found that *X. tulumensis* possesses venom glands containing digestive and paralytic toxins.[92] The initial observation that sparked the idea of researching the possible venomous nature of this remipede was that Dr. von Reumont and his colleague, Dr. Ronald Jenner, observed remipedes discarding empty shrimp shells out of their lairs. This inspired them to hypothesize that *X. tulumensis* preyed on shrimp, and that it may use venom as part of its hunting strategy. Since it is practically certain that we will discover many other examples of such venomous organisms in the future, some have proposed, rather speculatively for sure, that organisms like remipedes may be responsible for the unexpected (and sometimes mysterious) deaths of underwater cave divers,[93] which occur with some frequency and no apparent cause. Who knows?

» A SCORPION WITH TWO VENOMS

There is nothing new about scorpions being venomous.[94] In addition to the straightforward stinging behaviors that all true scorpions display, some species have developed quite sophisticated variations that, if nothing else, showcase nature's evolutionary ingenuity. I want to point your attention to a curious behavior evolved by at least one type of scorpion and another type of organism as well. This remarkable ability is the capacity to behaviorally modify their venom according to specific circumstances.

Parabuthus transvaalicus is a mean-looking scorpion, commonly known as the Transvaal thick- or fat-tailed scorpion, the dark scorpion, or the spitting scorpion. It lives mainly in southern Africa. As a regular guy, I am not particularly fond of scorpions or similar animals, but as a biologist I find them fascinating, in great part because I do so love venoms in general. As you know, venoms are wonderful biochemical examples of evolution in action. Their pharmacology and biochemistry are remarkable to be sure, but the wide variety of behaviors that animals employ to make use of such venoms is equally so. *P. transvaalicus* gives us an especially interesting example of such behaviors. These scorpions are sometimes called "spitting" scorpions for good reason. They can spray venom against a potential predator, with a fair degree of accuracy, too, and that is not even the coolest thing about them!

Have you ever heard the expression "to rub salt in the wound?" This refers to the quite painful sensation that one feels when salty substances come in contact with broken skin. It seems that the spitting scorpion has heard the expression too, and took full advantage of its wisdom. In a paper published in 2003,[95] scientists of the University of California at Davis reported that *P. transvaalicus* is able to produce a "prevenom" composed mainly of simple salts such as potassium chloride. Such salts are very easy to come by for any organism. In other words, they are metabolically cheap. When a predator, say a mouse, starts nibbling on the scorpion with the worst of intentions, the scorpion will initially sting the mouse, injecting it with the prevenom, this is, with mostly potassium chloride. More specifically, the scorpion is always "preloaded" with the prevenom. Such a salty solution will not likely kill the mouse, but it would sting like heck and will usually suffice to make the hungry mouse go away. This is in effect the equivalent of a warning shot.

If the mouse is *really* hungry though, it will continue biting the scorpion, and this continuous annoyance will trigger the synthesis of actual toxins, the real deal, you know, venom, as in "*no more Mr. Nice Guy.*" This two-step stinging process is undoubtedly a biological strategy bent on economy. From the perspective of the scorpion, it is much more expensive in metabolic terms to produce protein-based toxins, as opposed to making a simple salt solution. This makes sense, especially since these animals live in arid conditions, where resources are rather scarce.

This ability of an organism to make different venoms depending on the particular needs of the moment is not exclusive to scorpions. Some types of snails that we'll see in the next chapter (although we will not specifically talk too much about their venom) also produce different venoms according to specific circumstances.[96] I haven't heard of other examples of this strange ability, but I am absolutely sure that others will be found. Nature is quite thrifty and is always surprising us. All these variations, all these strategies, and all on only one planet! Can you imagine what could be "out there?"

» LEECHES

I may be wrong, but I do not think that many people have (or even consider having) leeches as pets, as these little guys are known heebie-jeebie generators. In fact, even the word "leech" triggers an icky feeling in most people, for they are the quintessential parasites: bloodsuckers. However, leech enthusiasts widely regard them as "*worms with* *character*."[97] You may be surprised to learn (as I was while researching the material for this chapter) that the original meaning of "leech" was "to heal," from the Anglo-Saxon word "*loece*." It seems that the first recorded use of the word "leech" in the English language dates from AD 900, and not surprisingly, the term was used in a medical context. It appeared in a book series called *Bald's Leechbook*,* which was a compendium of medical treatments and remedies as understood at the time. Pretty much since people began using this word, the relationship of leeches with the medical sciences was so close that, for a time, doctors (and even any kind of medical worker for that matter) were actually referred to as "leeches."[98] This is one of the ways we know that the medicinal properties of leeches have been known to humans for quite a long time. There are records from ancient Egyptian, Greek, Roman, and Byzantine physicians (among several other ancient cultures) describing some of their leech-related medical practices. As a matter of fact, some authors suspect that bloodletting as a medical practice has been around since the Stone Age. Leeches are even behind a rather well-known icon of modern society, the barber pole. Until relatively recently, the town's barber also served as surgeon or even dentist, depending on the specific needs of the customer. At any rate, they were in charge of bloodletting, which was standard procedure in those times, as ubiquitous as aspirin use is today. The bloodletting practice was rather systematic, and established practitioners usually had a series of essential materials available for the procedure: a rod or cane for the patient to grasp firmly to help the veins bulge, a container to either store the leeches or to catch excess blood, and, of course, pieces of cloth that were used as bandages or simply to clean up the mess. Over time, the bloodied bandages around the barber's rod alongside the leeches' washbowl became a symbol of the barber's practice, and this eventually evolved into the familiar red-and-white rotating pole.

---

* No relation to Yours Truly. (You see, I'm bald.)

The formal rationale for the practice of bloodletting stems in part from the master physician Hippocrates's humoral theory, which affirmed that good health was brought about by an optimal combination of the four humors present in the human body: blood, black bile, phlegm, and yellow bile. Some authors relate these humors to the four elements of antiquity: air, water, earth, and fire. Still others relate them to the four seasons.[99] We may nowadays chuckle a bit at this view of human health, but if you think about it, it is understandable. After all, the human body is capable of generating a few varieties of secretions, and in the absence of proper science, this was pretty much all that ancient physicians had to go on. It was pretty self-evident that when one or more of these fluids left the body in excessive amounts, one of the consequences was often death. In any case, the central idea was that good health was dependent on the proper balance of these four humors, and bloodletting was seen as a straightforward way to somehow bring them into balance.

Over time, bloodletting developed into an honored (and quite fashionable) medical practice, achieving such a level of popularity in the 1800s that the common medicinal leech (*Hirudo medicinalis*)* was becoming endangered. This particular species of leech is the one most commonly associated with human medicine.[100] In 1833, more than forty-one million leeches were imported into France alone! And they were quite expensive too, about fifty cents each in nineteenth-century dollars.[101] Keep in mind that corrected for inflation, etc., a nineteenth-century dollar was equivalent to about twenty-eight present-day dollars; therefore a single leech was worth approximately fourteen dollars. To give you some additional perspective, the total population of France in 1833 was about thirty-three

---

* There are several closely related species like *Hirudo verbana*, *H. provincialis*, *H. officinalis*, and *H. orientalis*. The "medicinalis" moniker in *H. medicinalis* is an indication of this species's close association with our medical practices. Incidentally, the word "hirudo" means "to cling." Very appropriate!

million![102] Yep, in the year of the Lord 1833, France used about one and a quarter leeches per French person.

Leech therapy is still going strong nowadays. Leeches are used to drain blood from wounds and to restore circulation in microsurgical procedures, among other applications, including plastic surgery. In fact, the leech is one of two animals that have the current approval of the Food and Drug Administration (FDA) to be used as medical devices. Incidentally, leech therapy was not limited to the medical sciences. Dentists have also taken advantage of the medicinal properties of leeches since at least the 1800s. They proved useful to deal with abscessed teeth and had the additional advantage of alleviating pain, which is probably the most feared side effect of dental procedures. I do not believe that leeches are still used in dentistry, but if you see a red-and-white striped pole outside your dentist's office, you might want to ask.

Leeches are formally classified as venomous, but the main purpose of their venom is not to kill prey or defend against a predator. Rather, leeches' venom is used for covertness and thievery. Their saliva contains a series of interesting compounds. Depending on the specific source and species, there are about one hundred chemicals, displaying a variety of physiological effects. Leeches "work" by latching onto the prey by biting the skin, then proceeding to extract blood. As you may imagine, a bite from such an organism will be noticed by the victim, which may well result in the emphatic detachment of the leech. From the worm's perspective, that will simply not do. Therefore, some of the compounds that the leech produces are intended to make sure that the prey does not notice the bite. Among these compounds we find anesthetic agents, which numb the affected area, and antihistamines, to attempt to prevent inflammatory reactions that normally occur at a bite site.

It is worth mentioning that most blood-sucking leeches are not able to directly obtain energy from feeding on blood. This digestive work belongs to a type of symbiotic bacteria (*Aeromonas hydrophila*)

within the leeches' gut, which help the worms digest the blood. Interestingly, this type of bacteria produces an antibiotic that prevents the
growth of other bacteria in the blood, thus preventing decomposition.
The bad news is that this species of bacteria is sometimes responsible
for infections in patients treated with leeches. This is something that
the physician (or barber, as the case may be) has to keep a close eye
on. At any rate, the chemical composition of leech venoms is truly
fascinating, which brings us to the point of our present conversation.
You see, there are several compounds in the venom of a few leech
species that interfere with the prey's blood coagulation events, with
the immediate benefit of enhancing blood flow for feeding. This is a
quite efficient system, as shown by the fact that some species are able
to ingest close to ten times their body weight in blood! Once the
leech has its fill, it simply lets go of the prey, but post-detachment,
the bite can ooze blood for several hours.* This is an indication of
the efficiency of the anticoagulant effect of the saliva. Of the various
leech substances that provide these anticoagulant properties, the best
known is a protein called *hirudin*. Undoubtedly this and many other
compounds are currently under close scrutiny in the hopes of finding
ways to develop and refine medical strategies against cardiovascular
and other types of diseases that involve (or utilize treatments involving) the process of coagulation.

» TOXIC BIRDS

Yes, there are such things as toxic birds. Depending on the specific
source and choice of classification, there seem to be at least fifteen
toxic/poisonous bird species. Many authors have expressed their surprise that this fact is not more widely known among the general

---

* I actually know of a case when this happened to a student who volunteered
to be bitten by a leech for a class demonstration. Quite a scare, but thankfully
the student (and the leech) lived.

public, let alone the zoological community. Part of this surprise is due to the fact that the toxic properties of certain birds have been known for quite a long time—even the ancient Greeks and Romans knew about some of them. Probably the most unexpected yet earliest known example would be a subspecies of the common European migratory quail. This is a generally edible bird that nonetheless expresses seasonal and geographical toxicity, meaning that it is safe to eat depending on the season and which direction the birds are heading. This is a puzzling and quite interesting story. This species of quail primarily follows three migration paths in its seasonal travels between Europe and Africa: On the so-called *western flyway* (Mediterranean Sea through Greece), the quail are poisonous during the spring (northward) migration but not during the fall (southward) migration. On the *eastern flyway* (Israel to southwestern Russia), the reverse is true: the quail are not poisonous while flying north in springtime, but they are poisonous while flying south in fall. And on the *central flyway*, where they fly across Italy, they are not poisonous in either direction. It certainly can become very confusing to figure out when it is safe to eat quail in the Old World!

I'll just have the chicken, please.

Because of the unusual pattern of toxicity displayed by these migratory quail, scientists believe that they acquire their toxicity from the environment, most likely from their food. This is not unprecedented—please recall the example of poison frogs. Moreover, we will see this type of toxin acquisition in other examples of toxic birds discussed in this chapter. The specific type of toxicity related to quail is termed *coturnism*, based on the genus of these birds (*Coturnix*), and its effects are characterized by abnormal muscle breakdown, with the expected associated consequences (eventual kidney failure, etc.). Not fun, to say the least. To date, the source or even the specific toxin or toxins that cause coturnism has not been identified. Some scholars have proposed hemlock seeds as the likely culprit, but the experimental data so far does not seem to support this idea. The jury, therefore, is still out on this matter. Interestingly, many scholars consider the biblical Book of Numbers (11:31–34 NIV) the earliest historical reference to the toxicity of these birds:

> 31 Now a wind went out from the Lord and drove quail in from the sea. It scattered them up to two cubits deep all around the camp, as far as a day's walk in any direction.

> 32 All that day and night and all the next day the people went out and gathered quail. No one gathered less than ten homers. Then they spread them out all around the camp.

> 33 But while the meat was still between their teeth and before it could be consumed, the anger of the Lord burned against the people, and he struck them with a severe plague.

> 34 Therefore the place was named Kibroth Hattaavah, because there they buried the people who had craved other food.

Even though it is evident that the toxicity of certain birds seems to have been common knowledge (for example, scientific hints about it have been commented on since the 1940s[103]), it was not until 1992 that scientists conclusively demonstrated the presence of a specific toxin in birds. These birds are the pitohui of New Guinea, of which there are several species, all of them toxic. Interestingly, an unrelated species, the blue-capped ifrit (*Ifrita kowaldi*), is also toxic. There is every indication that this is the proverbial tip of the iceberg because, in all probability, there are additional poisonous birds yet to be discovered. The current scientific consensus is that toxicity in birds is likely widespread. At any rate, the poisonous properties of the pitohui are well-known by New Guinea natives, who call one of the particular species (*Pitohui dichrous*, the hooded pitohui) the "rubbish" bird, on account of its inedible nature—in addition to its toxicity, the chemicals present in those birds make them quite unpalatable. Furthermore, if the bird is handled, it causes sneezing, nausea, and numbness of the mouth and nose. Also, if you have a small cut on your hands while touching it, it will sting like crazy. Just the smell of the bird is enough to make some sensitive people sick. These effects were what gave people a hint of the birds' poisonous qualities, prompting systematic research on them.

Scientists were in for a surprise.

One of the compounds present in these types of birds is not only quite toxic, it is exactly the same compound possessed by some poison frogs an ocean away. This toxin is called homobatrachotoxin, of which at least six varieties are known. The pitohui and related birds (and incidentally, the poison frogs as well) seem to obtain the toxins from their diet, particularly from a type of beetle called the *Melyrid* beetle. In general, the batrachotoxins are so nasty that they are considered some of the most potent nonproteins toxic to vertebrates (like ourselves). As an example, a batrachotoxin dose as small as 0.3 micrograms (μg—and just how small a microgram is will be discussed later in this chapter) is enough to kill a mouse in fifteen minutes or so. Translated to humans, a lethal dose would be about 0.75 milligrams (mg). In addition to the obvious defensive purpose for the presence of toxins in these birds, this toxicity also seems to protect the birds from certain species of lice and other parasites.[104]

There are several more examples of toxic birds, but the last one that I will talk about here is the spoor-winged goose. This bird is not exactly a goose, but it belongs to the same family, which is close enough for our purposes. This is an interesting bird in more than one sense. Its name comes from its possession of noticeable spurs on the wrists, which it uses for defensive and mating competition purposes. Its toxicity comes from ingesting a certain kind of beetle that produces the compound *cantharidin*. This is the main component of the fictitious aphrodisiac "Spanish Fly," but don't be fooled, this compound comes from the appropriately named "blister beetle," and it is an unmistakably toxic molecule. All these examples of apparently "acquired" toxicity by birds, courtesy of some insects, present a series of interesting questions. Perhaps the two main ones are: "How are the birds not poisoned by the chemicals produced by these insects?" and "How do the birds store and transport those toxic substances to specific parts of their anatomy?" These questions can also be applied to many other symbiotic relationships between organisms. There is so much to learn!

VENOMOUS MAMMALS*

Just imagine a fearsome mammalian predator, say, a lion, a wolf, or even a bear. I do not have to convince you that such creatures are, without a shred of a doubt, quite lethal, right? Now just imagine that in addition to the characteristics that make them dangerous, they are venomous as well.

Please do not fear, as there is no such animal in existence. Most predators in general do not need venom as part of their survival strategy—they are generally larger than their prey, and are therefore able to overpower said prey with sheer size and strength. Also, predators tend to be rather fast (although truth be told, some prey are faster!). Moreover, some big predators, for example lions and wolves, also have complex social structures that prove very useful for survival. At any rate, it just so happens that there are no venomous big mammals, thank goodness.[105]

As for small mammals, the reality is different. There are a small but fairly significant number of mammalian species that use venom for predation, defense, and in competition for mates. We generally do not immediately think "mammal" when we refer to something as "toxic" or "venomous," yet if we give the matter some thought it is not so surprising. Mammals did not start big in the history of life. They began as mouse- or shrewlike critters in the late Triassic period, close to two hundred million years ago, and stayed small for a long time, until the big rock that fell on Earth sixty-five million years

---

* For a very "reader-friendly" article on venomous mammals, please see Rode-Margono and Nekaris (2015).

ago wiped out most of the dinosaurs.* Anyway, mammals were small organisms for most of our early history. This meant, among other things, that their defensive capabilities were rather limited, namely being good at hiding, staying out of the way, and . . . not much more than that. However, there is evidence that points to the possibility of venomous capabilities in even some very early mammalian fossils. The first published evidence of anatomical features consistent with a venomous system in an ancient mammalian species dates from 2005. In this work, the investigators deduced the venomous properties of a sixty-million-year-old fossil that was a distant relative of the pangolins, currently living animals that are fascinating beings in their own right. The evidence that the researchers presented was directly related to the specific anatomy of the fossil's teeth, which displayed a series of grooves that are commonly interpreted as venom conduits. As we've seen, these "venom grooves" are a common characteristic in venomous animals. On the other hand, other authors argue that the first evidence of a venomous apparatus in mammals was published in 1988, with the description of a certain kind of spur in an ancient mammal, dating from the early Cretaceous period (more than sixty-five million years old). Modern examples of these spurs are found in the monotremes, which we'll talk about right now.

» PLATYPUSES AND ECHIDNAS

The first type of living venomous mammal that we'll discuss is the platypus; possibly its close cousin the echidna will prove to be venomous as well, but there is not enough evidence of that yet, so we'll limit this brief discussion to the platypus (*Ornithorynchus anatinus*). Both of these types of mammals are classified as monotremes, and I just have to say that monotremes are probably the weirdest mammals we have here on Earth, and weird in more than one sense. For

---

* Please note that I said "most" dinosaurs. Remember, there are living dinosaurs around—we know them as birds.

starters, they are the only mammals that lay eggs! They have additional oddities, such as a very avian-like bill, among other characteristics; regardless of whether they are venomous or not, they are not your typical mammal.

The platypuses are semiaquatic organisms that live in fresh water, mainly along the eastern coast of Australia. Only the males are venomous and even then they are venomous usually at certain times of the year, mainly during the mating season. In part because of this fact, the current scientific consensus is that they use their venom solely for defense and/or competition for mates. They do not use their venom to catch food; they feed on small invertebrates at the bottom of the pond/river. Most venomous vertebrates have their required anatomical structures inside their mouths (venom glands, fangs, etc.). Not the platypus. They have spurs located at their hind legs, roughly at the position where a thumb would be. Both sexes have them at birth, but the females lose them during development. In the platypus, the venom gland is called the crural gland,* which, together with the spur, makes up what is called the crural system. These spurs are strong bony structures. When disturbed, the platypus erects the spurs, wraps its hind legs around the attacker, and jams its spurs against the enemy. As stated above, the venom glands are generally active only during the mating season, so if the attack occurs during that period, venom will be injected. But even if there is no venom involved, it is a powerful jab, and no fun for the would-be assailant. The presence of such spurs in ancient mammalian fossils seems to indicate this was a common attribute in early mammals. The biological "purpose" of these spurs is less clear, as it is very difficult to establish toxicity exclusively by examining fossils. Nonetheless, it is within the realm of possibility that we mammals indeed started as venomous animals!

The platypuses' spurs are so strong that they will easily support the weight of the animal; when they have attacked humans,

---

* A technical term that simply means "belonging to the leg."

assistance has been oftentimes necessary to "detach" the platypus from its human victim. In humans, the venom causes quite a bit of pain—more like excruciating pain, according to most accounts—and swelling that can last for months. Moreover, the pain is usually insensitive to common anesthetics and analgesic agents. This suggests that their venom affects an alternate pain pathway, which is worth researching. On the positive side, there are no recorded human fatalities from platypus attacks, and in fact, there are only twenty-something recorded cases of human envenomation by platypuses. (They have killed canines, though, so don't bring one to a dog park!)

As mentioned, the toxic nature of crural gland secretions depends on the time of year when the attack occurs; it's more toxic during the mating season. Therefore, the severity of the sting ranges from just the unpleasant physical sensation of being gouged by spurs in the off-season, to a potent venomous assault during the mating season. The venom mixture has about twenty-odd components with various biochemical properties, all of them with close analogs in the venoms of other organisms. Genetic studies have concluded that there are close to one hundred venom gene candidates present in the platypus venom, which will undoubtedly be a source of some rather interesting molecules, with possible biomedical applications. With regard to the echidna, most scholars believe that they will be proven venomous, but the current research is inconclusive. An interesting point is that echidnas also possess spurs and proper crural glands, which produce a milky substance during mating season. I wouldn't be surprised if this secretion is found to have an important function as a pheromone* or even some defensive function. Only time will tell.

---

* A pheromone is a chemical compound made by an organism that serves as a signal to other organisms of the same species, usually for communication purposes (i.e., to attract prospective mates or to indicate the presence of an enemy).

» SHREWS AND SOLENODONS

Other types of mammals that display venomous properties are the shrews and solenodons, which are closely related. They are mostly insectivores, although they also eat other small invertebrates. These organisms, particularly the shrews, used to comprise the majority of venomous mammals known (not anymore; more on that later). These animals have a series of distinctive anatomical features, namely an elongated snout with venom glands located in the lower jaw. So far, there are three species of shrew proven beyond any doubt to be venomous: the North American *Blarina brevicauda* (Isn't it a pretty name? It is probably my favorite scientific name ever) and two European species that also live in parts of Asia: *Neomys anomalous* and *Neomys fodiens*. There are several other species "innocent until proven guilty" as far as venom-producing capacities are concerned. That being said, there is some evidence indicating that the Canarian shrew (*Crocidura canariensis*) can paralyze lizards upon biting them. If you ask me, it is only a matter of time before we have evidence of additional venomous shrews.

There is an interesting detail about *B. brevicauda* worth mentioning. This shrew displays a curious behavior, namely the "hoarding" of prey, meaning that it paralyzes such prey (small frogs, other small mammals, earthworms, etc.) and carries them—still alive but unable to move or otherwise defend themselves—to its lair for later consumption. This behavior is not uncommon. Certain species of moles have been observed to paralyze earthworms with their bite, and they also display this hoarding characteristic. Moreover, some insects famously hoard their prey!

The best-characterized venom-derived toxin found in shrews is called blarina toxin (guess which shrew it's from?). This toxin is mainly paralytic, yet it also causes respiratory depression and seizures in laboratory animals. In humans, the main effects are only swelling and pain, most likely because of the small amount of venom that an individual shrew can administer in a single bite.

Solenodons are relatively rare mammals whose natural distribution is limited to Cuba and Hispaniola (Haiti/Dominican Republic). There are only two extant* species: *Solenodon cubanus* and *Solenodon paradoxus*. There used to be a solenodon—albeit truly shrewlike—species in my native Puerto Rico (*Nesophontes edithae*), but they died out more than a thousand years ago. Because of their rarity, solenodons have not been as extensively studied as the shrews, but the best available evidence indicates that they are venomous as well. Sadly, as I mentioned, they are quite rare, and although currently protected, they are likely on their way to extinction.

STORIES, SPEARS, AND SERENDIPITOUS INDIGENOUS TOXINS

Recorded history is full of examples that implicitly or explicitly describe harmful, medicinal, and even quite psychoactive substances found in nature. In many cases, these substances or their specific sources were the origin of very interesting and curious legends. Some of these tales likely inspired concepts like "magical powers," and perhaps even stirred the beginnings of religious and spiritual practices as well.

One of my favorite examples of these legend-inspiring substances is the story of the *shark god* of the Hawaiian island of Maui. The legend tells the account of the capture and execution of one such shark god near the town of Hāna.[106] The story begins with a series of disappearances of fishermen from a certain village. Over time, the villagers associated these missing persons with the presence of an old hermit who lived on a cliff near the ocean. Upon capturing the old man, they discovered that he was not human at all, he was actually a shark god—the large mouth with big, menacing, triangular teeth on the hermit's back gave

---

* An extant species refers to a currently living species, as opposed to an extinct species, which is no longer around.

him away. This particular shark god got into the habit of eating fishermen and, quite understandably, the villagers did not like having such a monster in their midst. So they proceeded to kill the shark god, and just to make sure that he was really gone, they burned him and scattered his ashes in a nearby tide pool. The legend says that sometime after the event, an odd kind of seaweed began growing in the pool. The villagers named this seaweed *Limu-make-o-Hāna* (Hāna's deadly seaweed) because of its "magically" lethal properties. They noted that upon rubbing the tip of a spear on this seaweed, the spear seemed to acquire the power to rapidly kill anyone cut by it.

Fighters of this culture gained a reputation as fearsome warriors, quite possibly (at least in part) because of the unusually mortal nature of the wounds caused by their spears. Based on the shark god legend, it is easy to imagine a ceremony in which warriors would symbolically impale the *Limu-make-o-Hāna* with their spears, therefore endowing them with the seaweed's magic—hence, in their minds, securing their victory in battle.[*] Psychologically, *knowing*, as opposed to *hoping*, that their magic would work must have significantly boosted the warriors' confidence. Also, the possession of such magic would be an invaluable tactical asset; it stands to reason that they would have kept the source of that magic and its location a heavily guarded secret.

As frequently happens in folkloric accounts, there is a kernel of truth buried inside this legend. In the 1960s, scientists discovered a new species of *zoanthid* (a fancy name for certain types of corals) in a tide pool very much like the one in the *Limu-make-o-Hāna* legend. Scientists named this particular species of coral *Palythoa toxica*. The discovery of a new soft coral type was hardly remarkable; scientists had known of similar organisms

---

[*] This scene admittedly came from my overactive imagination. The reality seems to be just that the warriors simply rubbed their spears on the coral.

since at least the 1800s. The interesting detail about this particular coral is the later discovery, in the 1970s, of *palytoxin*, a nasty (and quite complex) substance produced by *P. toxica*, for which the toxin was named. Over time, scientists discovered that palytoxin was also found in other types of organisms, which is kind of surprising, given the molecular complexity of this toxin. All in all, the available evidence seemed to indicate that the deadly seaweed of legend was not a seaweed after all, but a class of coral, very much like *Palythoa toxica* and therefore that palytoxin, or a similar molecule, was the probable source of the *Limu-make-o-Hāna's* lethal "magical properties."

Unassumingly Lethal

Consistent with the *Limu-make-o-Hāna* legend, palytoxin has been shown to be dangerous to humans; even microgram amounts can prove lethal. You should be amazed at the previous sentence, because a microgram is a tiny, tiny amount! Let's take a moment to consider how little a microgram of anything really is. In order to do that, we can think about an object common in our daily experience, say a metal paper clip, a dime, or the cap of a ballpoint pen. Each of those weighs approximately one gram. If you divide that mass a thousandfold, you get a milligram (mg, as in the amount by which most medications are prescribed). Now, if you divide one of these milligrams by another thousandfold, you get a microgram (µg, or 0.000001 of a gram, which is pretty evidently a barely noticeable particle). The best available evidence seems to indicate that the lethal dose of palytoxin for a grown human is about five micrograms. This makes palytoxin one of the most potent toxins known—though believe it or not, there are toxins more potent still.

Palythoa corals are particularly beautiful and hence quite popular with saltwater aquarium hobbyists, and interestingly, there are several well-documented cases of palytoxin poisoning—

though not deaths—in aquarium enthusiasts. These unlucky individuals were likely poisoned when a small cut or scratch on their skin was exposed to the coral. The palytoxin story is no mere historical curiosity; it has practical implications as well!

For another example of the human use of harmful materials conveniently found in the environment, we travel to South America, where natives of this continent mastered the use of certain pharmacological agents to hunt and to wage war. Many tribes used a variety of plants for these purposes, including the extracts of the curare plant, *Chondrodendron tomentosum*. An extract of this plant applied to the tip of a blow dart makes it quite lethal. Throughout history, such darts were in fact dubbed the "flying death." Modern accounts of curare use date from the 1500s, when Sir Walter Raleigh described how South American natives prepared and used these plant extracts, generally known as *ourari*.[107] The common procedure entailed the collection of plant stems, "brewing" them in boiling water until reduced to a thick liquid. The natives then used this liquid to coat the darts' tips, which were ready when dry. When an animal is hit by a properly prepared curare-coated dart, it rapidly becomes paralyzed and dies within minutes, usually because of respiratory depression.

The curare plant extract contains a substance (d-tubocurarine), of which there are various forms. This substance causes paralysis and eventually respiratory failure by blocking a specific neuromuscular protein. The first formal experiments done with curare were performed by Sir Benjamin C. Brodie in 1811, but it was not until 1835 that d-tubocurarine was isolated in pure form by Dr. Harold King, using an old curare sample stored in a museum. The curare in this sample, remarkably, was still chemically active. This breakthrough was quickly followed by other work isolating several types of the molecule from a variety of specific sources. In the 1850s, the famous French physiologist Claude Bernard established the neuromuscular junction as the physiological target of curare. The neuromuscular junction is the point of contact between a nerve cell and a muscle cell; the former activates the latter. Bernard's discoveries, alongside the work of several other scientists, paved the way for the initial development of muscular pharmacology and physiology, which framed many of the subsequent discoveries in neuropharmacology. Later on, curare was used in the biomedical sciences, particularly in the development of chemical agents administered during anesthesia. In essence, curare made its way from the South American tropical forests to the operating room!

Curare is one of the earliest instances of a specific toxic chemical produced by nature that was identified from indigenous sources. In fact, some say that for ethnobotanists, curare represents a kind of Holy Grail that played a central role in the history of neuropharmacology.[108]

» BATS

There are three extant bloodsucking—or rather, blood-lapping—bat species: *Desmodus rotundus*, *Diphylla ecaudata*,* and *Diaemus youngi*. Blood-feeding bats are found solely in Central and South America. They are strictly classified as venomous as per the customary definition, but with a special twist. In contrast to most venomous animals, in which the venom is meant to produce an unambiguously noticeable result, namely to kill or repel, bat venom is used more like that of leeches, for stealth and stealing, since like the leech, it is against the bat's best interest to be noticed while feeding. Bat venom is a mixture of protein-cleaving and anticoagulant compounds. They inflict a small bite, mere millimeters across, and as the blood flows out of the victim, they gently lick it. There are no fatalities from this predation except in the case of smaller animals like chickens, which may bleed

---

* This particular species was recently reported to feed on humans. Please see Ito and collaborators (2016).

out due to the anticoagulant properties of the bat's venom, which can delay coagulation for several hours. By the way, the main anticoagulant protein in bats is called—wait for it . . . Draculin.

Draculin is currently being investigated as a medication source, and it is a safe bet that there are a few other vampire bat–derived compounds that will prove useful in biomedicine. In addition to the anticoagulants and protein-breaking compounds already discovered in bat venom, there is another interesting possibility: The victim of a vampire bat rarely notices the attack, which means that the skin may be numbed somehow. That side effect opens the door to discovery of novel anesthetic agents.[109] The future of this research is bound to be interesting!

Now let's move on to what is probably the most surprising example of a toxic or venomous mammal yet (at least until a living venomous saber-toothed tiger is discovered by an adventurous cryptozoologist!).*

» THE STRANGE CASE OF THE SLOW LORIS AND ITS ELBOWTOXINS

Recently, even a primate joined the list of venomous organisms. There are eight species of slow lorises, all belonging to the *Nycticebus* genus, and all of them seem to be venomous, making slow lorises the most abundant type of venomous mammal to date. Looking at a slow loris, you would never guess that it is such a dangerous animal. They have big, beautiful eyes and a facial expression of permanent melancholy, which makes you want to cuddle them up. Alas, however adorable they might appear, these are fairly aggressive creatures with a mean attitude and a really nasty bite.[110] Inside their mouth is a toothcomb made of sharp, needlelike teeth that is ostensibly used for grooming and feeding, but which we now think also aids in spreading their venom. In

---

* Not likely to happen, but I think we'd all agree that to see one of those would be really cool—from a safe distance, that is!

contrast to virtually all venomous reptiles or mammals, the slow loris does not have venom glands located in their mouth region. Rather, they have what scientists refer to as *brachial glands*, which means that their venom glands are located at "*relatively hair-free, slightly raised areas in the flexor region of the upper arms.*"[111] In other words, near their elbows. I am surprised that no one, so far as I know, refers to the loris's toxins as the "elbowtoxins." Maybe I am the first!

The loris is constantly licking its venom-gland secretion, which mixes liberally with its saliva. When it bites, this mix is what is "injected" into the would-be attacker. By the way, slow lorises have a curious way of advertising their dangerous nature. When threatened, they raise their arms above their heads, showing their brachial glands, similar to when a hero—or bad guy—in a movie pulls aside his coat to show off his gun.

There are several unusual aspects of the venomous behavior of lorises: First of all, we are dealing with a mammal here, and a primate, no less. What's more, the location of its venom glands is highly unique; they have no parallel in the animal world—no other mammal has venom glands near its elbows. In addition to these two remarkable facts, it seems that the slow loris is the only organism that produces a two-stage venom, and this seems to be related to their self-licking habit. This curious property of slow loris venom was discovered in a laboratory setting when testing the loris's glandular secretions in mice. When the venom from the glands was injected alone, it killed roughly 40 percent of the mice—not a weak venom effect by any stretch, but not particularly lethal, either. The researchers then decided to mimic the natural delivery of the venom by mixing it with the loris's saliva. When they administered the venom so mixed, 100 percent of the injected mice died; now that's lethal, all right! The data led investigators to believe that the venom is activated by some component in the saliva.

If humans are bitten by slow lorises, the bite is quite painful and has even resulted in fatalities. In these cases, the lethal quality of the venom seems to be related to a strong immune response called anaphylactic shock,* a phenomenon—not unique to loris venom, it also happens with bee stings, for example—that can indeed be fatal in sensitive individuals if untreated.[112]

The current understanding of the natural purpose of the loris's venom is that it is defensive as well as for competition with members of its own species. In fact, fights between members of the same species constitute the main risk factor for mortality in captive slow lorises. The lorises also coat themselves and their pups with the saliva-activated venom, which suggests an antiparasitic function. In fact, juvenile lorises produce their own venom and are actually more toxic than adults. This makes sense, because the young pups are more likely to be exposed to danger than adults of the species.

TOXINS AND VENOMS
IN THEIR NATURAL CONTEXT

We are very adept at taking advantage of the natural world and applying any knowledge that we obtain from nature to innovative purposes. For that reason, we sometimes forget that nature does not have us in mind when evolving. There are very specific contexts for each biological trait we care to examine, and just as with beauty, the nastiness of a substance is in the eye of the beholder. A very well-known example is an extremely toxic class of substances that include certain bacterial proteins collectively called *botulinum toxins*. You may have heard of one of these toxins, commonly referred to as botox. It is widely used in clinical settings as well as in some cosmetic procedures. Like curare, its mechanism of action involves the disruption

---

* There is an article that shows a dramatic example of an anaphylactic reaction to a slow loris's bite. The victim's face swelled up just like a cartoon character! The poor guy was treated and thankfully survived, and his face eventually came back to normal. Please see Madani and Nekaris (2014).

of neurotransmission from nerve to muscle, albeit through a different biochemical pathway. Also like curare, this toxin is particularly
dangerous when the affected muscles happen to be those in charge
of breathing. The lethal dose of botox for humans is about ten times
smaller than the lethal dose of palytoxin, which you will recall is a
mere five micrograms or so. This means that for botox, the lethal dose
for a grown human would be about *half* a microgram; this is barely a
speck of dust! Botulinum toxins are really nasty chemicals; they are
not to be messed with.

Despite botox's evident toxicity to humans—and in fact *because*
of it—we can actually use the example of botulinum toxins to illustrate a general point about the proper place of toxins in nature. You
see, botulinum toxin is not equally nasty to all creatures on this
Earth.[113] For example, there are many animals that survive by consuming carrion—sometimes simply roadkill—which by its nature
tends to be in various states of decay, and therefore it is likely full
of bacteria of all kinds, including the types of bacteria that produce
botulinum toxin. Remarkably, it is well documented that quite a few
classes of carrion feeders like vultures, for example, are resistant or
even immune to these types of toxins. In other words, we can imagine a hypothetical alternate universe, a universe where vultures possess a technological culture. In such a culture, their toxicology books
would not list botulinum toxin as an especially noxious toxin or even
as a toxin at all, for that matter. Botox may be portrayed in their
cookbooks as a condiment instead, and they might sing the praises of
properly aged roadkill!

The point that I want to make with this example is that there is
a proper *ecological context* that we must take into consideration when
describing a toxin candidate.[114] Such a description ideally would
include an explanation of its possible mechanism of action in light of
the toxin's likely target organisms. On the other hand, we know that
all life on Earth is related. This similarity makes humans sensitive to
many toxins "intended" to kill other species. For example, it is very

unlikely that we humans are the food source of rattlesnakes, wasps, or even Australian funnel-web spiders, but their venom can be fatal to us all the same.

I barely scratched the surface of the topic of venoms and toxins in this chapter; there is so much more to know! There are quite a few general books that cover this fascinating area of the biological sciences; please seek them out if I have piqued your interest![115] For now, let's continue our exploration of unique and quite unusual ways that living organisms compete in the game of life. In the next chapter, we'll explore a series of intriguing survival strategies with some common themes, namely surprise attacks with decisive speed and force.

CHAPTER 5

THE FAST AND THE HANGRY*

E verything moves. Everything in the universe is constantly moving, and I mean *everything . . . all the time.*** Even in your moments of true relaxation, you are moving faster than you can perceive, realize, or even imagine. At the "macro" level, you are on the surface of a planet that is spinning at an average velocity of about one thousand miles per hour, depending on the exact point on Earth you occupy at any particular moment.*** In addition to its spin, the Earth orbits the

---

* This happens when someone—or something—is so hungry that hunger leads to anger, frustration, and oftentimes both.

** This is hardly a modern idea; the Greek philosophers thought about this extensively. One of them, Heraclitus (near 500 BC), formalized the idea that physical change is one of the constant properties in the universe, which he expressed as "Everything changes, nothing stays still . . ." This idea was particularly insightful, especially since many of the dynamic, ever-changing aspects of the universe—namely, the astronomical realm and the microscopic realm—were essentially unknown to the ancient Greeks.

*** Right at the equator, your speed would be exactly 1,036 miles per hour. Since we live on a sphere, the farther you go away from the equator—north or south—the slower the speed of spin, until you arrive at either pole, where

sun at about sixty-five thousand miles per hour. We do not feel any of
these movements because gravity binds us to the planet. Now imagine
going beyond our solar system. What comes next is our local group of
stars, then the whole galaxy, our family of galaxies, and so on. All of
them (and us) are moving through the universe at speeds far removed
from our daily experience.

Furthermore, the universe itself is moving, in an ever-going
expansion that began about fourteen billion years ago, an expansion
that paradoxically seems to be speeding up, but let's go back home
by zooming back to you and the comfy chair that you are probably
resting in. Even though you now know that you are moving, you do
not feel like you are moving at all—in part because you are no doubt
enthralled by the indescribable bliss that is reading this book. At
any rate, there is quite a bit of action going on inside your body at
this very second as well. Many of your muscles are busily contract-
ing and relaxing—again, outside of your sphere of awareness. You
are not consciously controlling them and you do not have to; they
know what to do and when to do it. Your heart is beating, pump-
ing blood throughout your body, and your diaphragm relaxes and
contracts, controlling the rhythm of your breathing. There are many
other things happening in virtually every organ system of your body
at this very second; you get the idea. If we zoom in even closer, at
the "micro" level, your tissues are in constant motion as well, and at
the molecular level, inside your very cells, there is a buzz of activity
that shames even the angriest hornet's nest that has ever existed. The
magnitude of these molecular movements is what gives a group of
such molecules its very temperature. We can continue zooming in
or out, at the macro or micro level, and we'd keep finding significant
movement activity, but I think that I have made my point. The universe never stands still.[116] And this is especially true of the biological
world.

---

your relative velocity is theoretically zero. But since not very many people live
exactly at the poles, we can ignore this fact for the purposes of our story.

THE ONLY REASON FOR TIME IS SO EVERYTHING DOESN'T HAPPEN AT ONCE*

It is very difficult to talk about movement without taking into consideration the concept of time scales. Time is seamlessly interwoven with biology. On the one hand, our brains can understand in an abstract way time periods like thousands or millions of years, or at the other extreme, milliseconds or microseconds. On the other hand, we can truly experience and therefore directly understand a rather limited set of such magnitudes, like seconds, minutes, hours, and days, for example. In the grand scheme of things, meaning geological and evolutionary time, events happen over thousands, even millions of years.** Whereas at the molecular level, biological systems work in time frames of milliseconds (1/1,000 of a second), microseconds (1/1,000,000 of a second), or perhaps nanoseconds (1/1,000,000,000 of a second). Certain chemical reactions can even be faster than this. We cannot, by any stretch of the imagination, perceive any of those time frames directly, but we can apply perspective. Please consider a common figure of speech that is meant to describe a very fast occurrence: "*in the blink of an eye.*" An eyeblink is a fast event all right; it is rather fast from a normal human perspective, as your usual eyeblink lasts for about three hundred milliseconds or so. However, just to give you some additional comparisons, in a literal eyeblink, the fastest hummingbird beats its wings about twenty-four times, and in that exact same period of time, a typical nerve cell fires almost exactly three hundred times. This brings us to a concept that implies a rather fast event: thought itself.

We are all under the illusion that we live our lives in real time. For pretty much all our daily activities, and as long as nothing happens beyond or below a certain speed, this is true enough. The oft-used phrase "the speed of thought"[117] represents how fast we think

---

* Attributed to Albert Einstein (probably apocryphal, though).
** Life itself has been evolving for the past 3.8 billion years or so, a number of years that we can articulate, but that we cannot truly comprehend.

thought is; we mainly perceive our thoughts as instantaneous, and for all practical purposes, we can say that this is true. However, the speed of thought is finite, and not particularly fast once we examine it more closely. Thought is generated by nerve transmission. Most estimates of nerve-signaling velocity range between 150 and 270 miles per hour. This speed, for all intents and purposes, is instantaneous—again, from our perspective. Alas, it is not fast enough to protect us from certain dangers. Say, for example, that an early human ancestor grabbed a hot piece of wood from a fire. If she had to rely on her conscious mind to notice and react to the heat, she would likely get badly burned.

Thankfully, what most likely happened then (and still happens today) is that even before your ancestor was conscious of the temperature of the piece of wood, a reflex arc went from a sensory neuron to the spinal cord, then to another neuron, and then to a motor neuron, which finally contracted the muscles that made her let go of the stick immediately, minimizing the skin damage.

I know that the previous sentence was quite long, but I wrote it like that on purpose. You see, amazingly, this series of events happens at a breakneck pace—coincidentally, at about the same speed as an eyeblink. The sensory message that made her feel the temperature of the piece of wood did not even reach her brain until later, much later, after letting go of the wood. In a literal sense, she did not have to feel the burn on her skin for her body to react to it. Keep in mind that every sensation may come from say, an extremity, but you feel it in your brain, the only organ capable of decoding sensory stimuli and translating them into mental sensations. The main message here is that there are things faster than thought. Think about that for a moment.

BIOLOGICAL BALLISTICS

Biological organisms are made of matter, and are therefore subject to the rules of physics and chemistry. We must obey the universe's physical laws, meaning, among many other things, that we cannot travel faster than light, that gravitational fields give our mass the impression of weight, and that the energy we can carry and transfer at any given time depends on how fast we are moving. This last one is best known as Newton's* second law of motion, which essentially means that the force that any given amount of matter can exert depends on the rate of movement per unit of time; in other words, its acceleration. In short, the heavier an object is, and the faster it goes, the greater the punch it can deliver.

A consequence of Isaac Newton's second law is that deceptively small objects that carry a lot of energy can cause a lot of damage, if they are moving fast enough. A bullet, for example, is not a very heavy object, as an average bullet weighs about eight grams (about one-third of an ounce).** This is not a lot of weight. And yet, without going into too much detail, even such a small piece of metal can be lethal because of the acceleration imparted by the ignition of a certain amount of gunpowder. Or think about the asteroid or comet that signaled the end of the dinosaur era (briefly mentioned earlier in this book). This big rock that crashed on our planet was about six miles long (think about this: a car moving at sixty miles per hour takes six minutes to cover six miles). Sure, that is big by human standards, but to get a real perspective of its size compared to the Earth's, just look at a world globe or a world map and figure out approximately how much surface area anything six miles long would cover. In fact, it covers a minute fraction, almost invisible. Surprised? Yes, from this planetary perspective, that rock was a rather small object indeed, a mere pebble compared to the size of our planet. However, that "mere pebble" fell on us at a speed of about *forty thousand* miles per hour! That cosmic piece of dust triggered a series of events that changed

---

* After Sir Isaac Newton, the scholar and polymath who gave us the first scientific theory of gravitation, among many other things; he also gave us an entirely new math, called calculus (almost simultaneously with the German philosopher Gottfried Wilhelm Leibniz).

** In comparison, a metal paper clip weighs about one gram, a US quarter weighs almost six grams, and so forth.

the Earth and its biosphere forever. That rock packed such a punch that today—sixty-five million years after the event—there are still traces of the crater that it left on Earth.*

Seemingly small things can indeed amount to quite a lot depending on the specific situation; one factor is how fast they are moving, which influences how much power is released in a certain period of time. And this does not apply just to planetary events or other non-living systems. Biology has for the longest time excelled at applied physics; there is no question about that. Many organisms have evolved strategies to take full advantage of our physical world. Not surprisingly, most of the activities that reflect these strategies directly contribute to the undertakings that we know are nonnegotiable in order to stay alive: obtaining nourishment, avoiding becoming nourishment for others, and reproducing.

One of the most majestic technological displays of our current mastery of physics and mathematics is a rocket launch. I've never seen one—though if I ever make a bucket list, a rocket launch will surely be on it—but for now I rely on the eyewitness accounts of hundreds of people, who attest to the awe-inspiring power of such an event.

Emotions aside, we can learn a lot about biological survival tactics by observing some of the characteristics of a typical rocket launch. The main point, of course, is to escape Earth's gravity and get into orbit. To do that, a vessel must reach a minimum velocity, formally called *escape velocity*, which is the absolute minimum speed that an—upwardly flying—object must reach in order not to fall back to the ground. For our planet, that velocity is about 11 km/s, or about 25,000 miles per hour (the bigger the celestial body, the larger

---

* It is called the Chicxulub crater, located beneath the Yucatán Peninsula in Mexico. You cannot see the crater directly. To start with, it is big, about one hundred miles long. Also it is buried, so it is not readily visible. The crater remnants were discovered in 1978 by two geophysicists, Antonio Camargo and Glen Penfield, who were prospecting for petroleum for an oil company.

the escape velocity). Once in flight, the astronauts inside the rocket are subject to increased g-forces; 1 g equals the gravitational attraction of the Earth (again, the bigger the planet the higher the gravity; this is why on this planet you may weigh two hundred pounds yet your weight on the moon—with no change in mass—would be about thirty-three pounds). Astronauts on their way up are exposed to g-forces about three times higher than on the Earth's surface. It may not sound like much, but this means that at that particular time, a two-hundred-pound person feels like—and literally weighs—some six hundred pounds.

Well, certain living organisms are capable of achieving higher apparent accelerations and therefore greater gravitational forces at much smaller scales. From fungi to plants and jellyfish, many of these organisms somehow generate tremendous forces that without a doubt put to shame the fastest contraptions ever made by people. A case in point is the jellyfish sting.

» CNIDARIANS

When a jellyfish—or a related animal, like some corals, sea anemones, and even the hydra that we saw in an earlier chapter—stings you, likely the thought farthest from your mind is the exquisite mastery of biomechanics that allows such an entity to inject its venom into its victim, in this case, you. For reasons that will be immediately apparent, jellyfish, corals, sea anemones, and hydra are classified in the same animal group: the cnidarians. All of them are venomous, and depending on the specific kind of organism, the sting may be only somewhat annoying or amazingly lethal. With rockets and big speeding asteroids already in our minds, let's talk about how amazingly fast these small organisms are able to sting and how they are able to do it. (We'll not go into the specifics of cnidarian venom here; as fascinating as they are, they have been extensively reviewed elsewhere.[118])

Have you seen—in pretty much any good science-fiction war movie—small enemy ships swarm and crash into big cruisers at high speed, piercing the hull and letting the invading soldiers, well, invade?* Every time *I* see a scene like that, I think of jellyfish. Do you want to know why?

Jellyfish and related organisms have a very special kind of cell, one that is a true biological weapon. These are called either *nematocysts* or, more recently, *cnidocysts*. It seems that they have been observed since at least the 1700s, but their specific function was not discovered until the nineteenth century. Cnidocysts are highly specialized cells, found only in cnidarians, and in fact these cells are the basis for the classification of these organisms, since as far as we know, cnidocysts are not found in any others. Cnidocysts come in a variety of sizes—from 3 μm to about 100 μm—and display quite a few different shapes. There are about thirty distinct types described to date, but they all share a fundamental structural feature. They have a venom-filled internal cavity that ranges from roughly cylindrical to almost spherical in shape. This cavity contains an attached hollow filament, coiled to fit within it. This filament ends in a dart- or harpoon-like tip, about 15 nm (nanometers) wide, which is frequently adorned with barbs and other structures. When the cnidocyst fires, the harpoon tip is launched at very high speeds, allowing it to penetrate the target's skin, piercing it. Once the harpoon pierces the victim's skin, the harpoon opens up and turns itself inside out, allowing the filament itself to enter the victim's tissues and discharge the venom. Hence the comparison with a raider spaceship: the dart is like the tiny attack ship and the filament/venom corresponds to the invading soldiers. The filament remains attached to the insides of the target's tissues.

Cnidocysts are highly sensitive cells. They discharge through a variety of stimuli, including pressure (i.e., touch), chemical exposure, and in some cases, light of certain wavelengths. Talk about trigger-happy

---

* A visually stunning portrayal of this space warfare tactic is shown in the 2016 motion picture *Star Trek Beyond*.

organisms! This is an all-or-nothing event; once it starts, it does not stop; it has to run its course. Also, they can fire only once; they cannot be reused. These apparent disadvantages are more than offset by the immense number of cnidocysts in these organisms. They are, without a doubt, formidable weapons. Just like an attack fleet, one raider will do little, but millions of them at the same time will do a lot of damage.

Furthermore, cnidocysts are widely considered some of the fastest cells in the animal kingdom, and this is precisely the key that explains how such small structures, when fired, are able to break the tissues of the organisms' prey. How do they do it? Scientists have known for a while that cnidocysts are capable of generating high pressure levels within their internal cavity. It was evident that this was a very important factor. However, it was only about ten years ago that careful studies to measure the actual speed of discharge were undertaken. Previously, there were no camera systems fast enough to capture the process; a plain and simple matter of unavailable technology delayed the research. Using newly developed imaging techniques, researchers made a series of interesting observations about the cnidocyst discharge mechanism of a particular jellyfish. They found that the average time that passed between stimulus and discharge was about seven nanoseconds (this is about 0.0000007 seconds; please recall that a literal eyeblink takes about 0.3 seconds!). If a cnidocyst were able to fire multiple times, it would be able to fire 428,572 times in an eyeblink! Once triggered, the cnidocyst's harpoon travels over a distance of about 13 micrometers in about one microsecond. Based on these numbers, scientists were able to calculate that the harpoon is exposed to a gravitational force equivalent to five *million* g's.* Astronauts on their way into orbit endure all of 3 g's or maybe a little bit more. A cursory calculation shows that the acceleration imposed on these stingers allows them to strike their targets at pressures comparable to the ones experienced by a gunpowder-propelled bullet. Though

---

* Yes, I double-checked their numbers.

cnidocysts' stingers are tiny, they are fired at a hard-to-comprehend speed that allows them to pack a powerful punch. I like to imagine our friend Isaac Newton smiling over these wee stingers, as the forces that result from their mass and their acceleration allow them to penetrate their victim's skin and release their venom.[119]

There are many other examples of "natural ballistics."[120] We have barely scratched the surface of high-speed biology! Let's continue our journey by looking at an especially striking example of a group of animals that takes full advantage of physics, particularly the physics of speed: these organisms are collectively called mantis shrimp, and we'll talk about them next.

LOOTERS AND PILLAGERS

You might be wondering why a creature like a jellyfish *needs* such a forceful and speedy weapon. The cnidocysts' harpoons are microscopic and therefore very light; to us, they weigh almost nothing. In order for them to penetrate any kind of skin, they have to compensate for their lack of mass by utilizing speed. Think of the example of a bullet. If someone throws a bullet by hand, the worst-case scenario for the living target is a nasty bump. Much more power, i.e., speed, is needed for the bullet to do damage, and that's where the powder and shooting mechanism, etc., come in.

As creatures of the ocean, cnidarians are both prey and predator. In their inherent quest to survive, some unusual organisms have found ways to avoid getting stung by cnidocysts, and these elusive organisms also actively feed on cnidarians. Furthermore, some of these predators steal the cnidocysts from their cnidarian prey and use them for their own purposes, meaning that the predator "installs" the stolen cnidocysts in their own skin. As far as we know, there are only three kinds of animals able to achieve such feats: one is ctenophores, which are jellyfish-like sea creatures that do not sting and in fact belong to a different group of animals. Then we have the nudibranchs, which are essentially sea slugs (and include some of the most beautiful organisms in the sea). Finally, we have our good friends the flatworms, especially a particular type of worms called polyclads.

It is remarkable to note that animals like these are not only immune to cnidarian venoms but are also able to manipulate cnidocysts without triggering their delicate firing mechanisms. In the case of some nudibranchs, there is research indicating that they seem to produce a kind of slime that somehow inactivates the stolen cnidocysts and protects the thief from their sting. This suggests a component of the slime may pharmacologically inactivate the firing mechanism, but not much is known about how exactly nudibranchs are able to pull this off, or even how they repurpose the stolen cnidocysts for their own defense. In the cases of the ctenophores and the polyclad flatworms, we know even less. How do they become immune to cnidarian venom? How do they deactivate cnidocysts to manipulate them? How do they reactivate them? Answers to these questions are especially important since jellyfish stings are a significant public health issue in certain countries. Sting-resistant animals will undoubtedly help us in this matter, and I fully expect quite a few surprises from the study of these interesting creatures.[121]

» MANTIS SHRIMP

Mantis shrimp are gorgeous and quite interesting animals, but they are not shrimp and they are most definitively not praying mantises. There are close to four hundred species of mantis shrimp worldwide, many of them beautifully colored. They belong to their own group, the *stomatopoda* (colloquially called *stomatopods*), but truth be told, some of them really do have a very shrimp- or lobster-like appearance. They do not tend to be very big; they range from about one to sixteen inches. They are ancient organisms with a distinguished paleontological history; their fossil records go back some three hundred million years.

These organisms all have a pair of appendages that closely resemble the praying mantis's front legs, hence the "mantis" moniker. Mantis shrimp do tend to be very aggressive, and they have the brawn to back it up. They also display quite a bit of anatomical variability, but they can all be broadly classified as either "spearers" or "smashers." These two main varieties display the remarkable predatory prowess of stomatopods via their mastery of speed, making excellent use of the physics of speed to procure their mealtime prey. Both spearers and smashers take advantage of the fact that any given energy they are capable of expending will be amplified by the time scale at which that energy is exerted—that is, how fast it travels. For example,[122] if you try to throw a stick using just your arm, the stick will travel some distance, but no matter how strong you are, the biology of muscle limits the speed at which you may throw the aforementioned stick. On the other hand, if you use a bow, the stick is no mere stick anymore, it becomes an arrow that will travel significantly faster and, upon reaching its intended target, may actually enter that target. This is another aspect of the mass/acceleration relationship. We can go further; think of a crossbow, which would be able to throw the arrow much faster than a normal bow, with significantly higher lethality.*

If we compare the time periods in which the arrow travels to its target—by the arm alone, or by bow, or by crossbow—we'll find that the same force exerted over shorter time periods produces significantly higher power amplification. This is where mantis shrimp excel. Their classification as spearers or smashers not only offers a clue about their specific prey, it's based upon the structure of their specific mouthparts called *raptorial appendages*. These appendages display a wide variety of shapes depending on the stomatopod species. What they all have in common is a mechanism that endows

---

* In fact, for a long time, crossbows were so feared that in AD 1139, Pope Innocent II banned the use of crossbows . . . but only against Christians. More than eight hundred years have passed since then, and for the most part we have not made a lot of progress in our treatment of people who do not share our beliefs; don't you agree?

these body parts with the capacity for high-speed movement. Their secret is a spring-loaded mechanism that acts in very much the same way as a bow or a crossbow, achieving speeds of about one hundred feet per second. This means that they can release a certain amount of force over a very short time period—in the order of milliseconds—effectively producing a high amount of power.

Spearers are the most ancient type of mantis shrimp and represent most species of this type of crustacean. Spearers are what one would expect: ambush predators that prey on soft and relatively fast creatures, such as fish. Their raptorial appendages end in sharp spinelike structures that they use in combination with the aforementioned high speeds to impale and trap the fish. Here is the part of the story that I like to call "the shrimp and the shock wave." About fifty million years ago, an alternate strategy evolved in certain species of mantis shrimp, leading to the rise of the much faster "smashers." This class of mantis shrimp preys on much slower targets, like snails and crabs. You would expect that the "faster" shrimp would go after faster prey, but in fact, the spearers are as fast as they need to be, as their trigger speed is oftentimes faster than the reaction speed of a fish. On the other hand, the smashers prey on slower—but much *harder*—victims, namely snails and crabs with shells. Smashers do not impale their prey; they break them. Smashers conserved and improved upon the ultrafast mechanism of the spearers for prey capture, and developed a series of shape variations in their appendages that resulted in a hammer-like shape of their raptorial appendages. Their strategy is simply to literally smash their way through the hard shell of their prey. In about a millisecond, they deliver a blow that, when well placed, not only cracks the shell, but is so fast that it generates a phenomenon called *cavitation*. This is a serious show of force. Mantis shrimp–induced cavitation occurs when the ultrafast movement of a raptorial appendage causes the water surrounding it to move so rapidly that it creates an air bubble composed mainly of water vapor; this bubble collapses at high speed, emitting heat and light as well as a high level of pressure. The end result is that the shrimp's original punch is immediately followed by a high-powered shock wave, the ultimate one-two punch.[123] Just to give you an example of how powerful a process cavitation is, it is a big problem for boat propellers. The cavitation generated by the high-speed movement of the propellers damages the metal over time. Interestingly, cavitation does not seem to be a problem for the stomatopod, and does not appear to affect the shrimp itself. In other words, the biological material of the animal's appendage does not get cracked or damaged.[124] Obviously, this represents an interesting line of research pertaining to the development and engineering of extra-durable materials.[125] The study of nature, yet again, may provide us with solutions to a wide variety of material engineering problems.

COLOR VISION

I find it quite interesting (and I hope that you'll share my fascination) that in addition to their mastery of the physical principles that endow them with the capacity for speed and force, mantis shrimp also have mastered the fine points of some aspects of the physics of light itself. Thus, I'd like to say a few words about another extraordinary characteristic of these organisms: their highly unusual visual system.

Even though we perceive the world around us through the signals that reach the brain via our usual sensory modalities (sight, taste, smell, touch, and hearing, among several less well-known others), we humans are overwhelmingly visual organisms, barring disease or accident. Our interpretation of sight is brought about by the detection of photons, which are essentially light particles.* Photons come in a wide variety of wavelengths, from the very short (as in gamma rays) all the way to the very long (as in radio waves), organized as the so-called *electromagnetic spectrum*. We

---

* Photons are also waves, but we do not have to concern ourselves with the physical basis of this fact here except to say that we can talk about light in terms of wavelength.

have the means to detect virtually any of these photons with the appropriate contraptions and techniques. However, humans are only able to see by ourselves a very limited slice of the electromagnetic spectrum. Not very imaginatively, we name the part of the spectrum that we can see with our naked eyes "visible light." It is well known that organisms like some insects and some viper-like snakes are able to detect and image photons in the ultraviolet and infrared wavelengths, respectively. The world certainly looks different to these creatures.

Even though the visible part of the spectrum is a mere fraction of the whole thing, under normal circumstances humans can distinguish roughly one million colors. How is this possible?

Well, humans and other organisms detect light via a series of specialized cells called photoreceptors.[126] These cells come in two main types. The first are the rods—named so because of their cylindrical shape—which work best in dimmer light, as they are specialized to detect light intensity, as opposed to color per se. There is just one type of rod, which has only one specific type of photoreceptive (light-collecting) pigment. Therefore, rods are not very good at detecting colors. For example, when you are in a dark room, you may be able to see the shapes of objects around you but not their colors. In such a circumstance, you are mainly using your rods.

Then there is the other variety of photoreceptor cell in our eyes. These are called cones (again, on account of their overall shape), and this is where color perception comes in. Humans have three types of cones, each one specialized to detect a limited number of light wavelengths, roughly red, green, and blue. Each cone is capable of detecting close to one hundred different wavelengths, give or take. Since we have three distinct cones, all the possible combinations that can be detected, integrated, and sent to the brain for perception are about 100 x 100 x 100 = 1,000,000. We call our type of visual system a *trichromatic*—as in three basic colors—system.

By now, you should know that nature is rarely simplistic. About 12 percent of the human population develops an additional type of cone, with its own wavelength sensitivity—interestingly, most of these individuals are women. We refer to such individuals as possessing *tetrachromatic* vision ("tetra" meaning four, as in four cone types). This additional cone provides a thousandfold capacity to perceive various wavelength combinations. Therefore, they can in theory perceive 100 x 100 x 100 x 100 = 100,000,000 colors!

As impressive as tetrachromatic humans are, nature presents us with some organisms with even more staggering visual capacities, like—you guessed it!—mantis shrimp. Their visual system seems to work a little bit differently from ours,[127] but the main point is that they have possibly the highest number of photoreceptor types of any organism ever studied. They have twelve types of cones, which in itself is impressive; as discussed above, a tetrachromat can potentially see about one hundred million colors. It is rather straightforward to calculate how many approximate colors the typical mantis shrimp can see—there will be quite a few zeroes in such a number! In addition to these twelve photoreceptor types, most species of mantis shrimp also have photoreceptors for a few types of polarized light. We humans simply cannot see polarized light without the help of specialized materials. The capacity to perceive polarized light provides these types of mantis shrimp with previously unknown visual capacities.[128] Can you imagine the kind of art that a mantis shrimp version of Vincent van Gogh would be capable of creating?

» KILLER SNAILS

Cone snails include some eight hundred species distributed worldwide. This is the largest genus of marine animals overall! They comprise a rather young group, evolutionarily speaking. It seems that they are no more than fifty million years old or so. And yet, very few animals have captured our attention like the cone snails. At different points throughout history, they have been sought after with various motivations: for their aesthetic appeal, their unique biological characteristics, their behavioral traits, and their importance to the biomedical sciences, among many others.

In the spirit of full disclosure, I would like to say that these organisms are very close to my heart. It is kind of a sentimental matter. My very first scientific paper, which I coauthored with two of my scientific mentors, was about a special characteristic of the venom of a species of certain killer snails.[129] Simply because I like these snails so much, this section will be slightly longer than the other sections in the chapter. Would you please forgive me this indulgence? It'll be worth it—you'll see!

As mentioned in an earlier chapter, when asked to think about a venomous or toxic animal, most of us likely think of the usual suspects: snakes, scorpions and spiders, wasps and bees, etc. However, there are many unusual suspects, as we saw in the chapter by that title. In any case, it would be very unlikely that anyone's first thought would be of snails. You may be surprised to hear that certain species of snails produce toxic molecules and can therefore be unambiguously called venomous. This is not all that far-fetched. Snails are not only usually small but also very slow animals. Being toxic would help them to avoid predators. We should not be too surprised to find that some snail species use toxins as a defense mechanism. Possibly even more interesting are the behavioral strategies that some of these snails have evolved that go beyond the simple delivery of poisonous substances. Some of these snails possess a degree of behavioral sophistication that most of us admittedly biased vertebrates would not expect from invertebrate organisms.[130] What if I told you that there are some marine snails that are not only venomous but are also active *hunters*?

Yes, there are thousands of species of hunting snails, all of them marine, and some species of these snails exclusively hunt fish! Now, this is quite surprising since no snail is known for its visual acuity, agility, or its speed, all traits that are rather useful for hunting, especially when hunting fish. You likely know that fish can be quite skittish—ask anyone who has ever owned a fish tank. One merely needs to wave a hand over a fish tank to induce a behavior called the *startle response* or *startle reflex*, which is exactly what it sounds like. When the startle response occurs, the fish rapidly swim everywhere and anywhere but the place they were before. This is a rather useful defense mechanism that depends on the alertness, reaction time, and speed of the fish.

This common fish behavior begs the question: How would a slow snail be able to catch fast fish? The answer seems to be a combination of four main aspects: stealth, venom, sheer natural cleverness, and surprisingly, speed. There are at least three broad types of marine snails that display active hunting behavior: these are the *turrids*, the *auger*, and the *conus* snails. These three types of snails catch their prey by injecting them with complex venoms through a rather sophisticated "harpoon," which is a modified tooth that acts very much as a hypodermic needle. Each species displays a specific tooth shape and size; this feature is one of the criteria used to classify these organisms. In fact, many authors state that harpoon morphology is enough to identify a given species. The auger and turrid snails are mainly worm hunters; these snails are generally small and relatively lesser studied. However, they are currently under close scrutiny as they also show a great deal of pharmacological promise because of the complexity of their venom. The toxins made by cone snails are collectively called (what else?) *conotoxins*. The cone snail's prey is paralyzed so quickly because the venom contains a number of specific toxins that target various proteins controlling the nervous system activity of the fish. "Modern" cone snail research began in the 1970s with the work of Dr. Baldomero M. Olivera and his collaborators at the University of Utah. His story is an example that illustrates how many apparently unconnected events can coalesce and lead to major discoveries.[*]

---

[*] As remarkable as Dr. Olivera's story is, he stood on the oftentimes mentioned "shoulders of giants" to expand upon this exciting area of research. He was not the first scientist captivated by these snails; he was not even the first one to study their venoms. Interestingly, most likely one of the reasons

Baldomero Olivera was familiar with cone snails from child-hood. In his native Philippines, he collected shells, including the brightly colored cone snail shells, and he knew the stories that told of how poisonous these snails could be. (Incidentally, cone snail soup—sans venom glands—is a delicacy in the Philippines.) Life and schooling interrupted his hobby, for a short while at least.* Years later, Dr. Olivera returned to the Philippines with the intention of establishing his own research program in molecular biology there, but a relative lack of funding and equipment prompted him to look for alternate projects that he could undertake in his native coun-try. He remembered his childhood hobby and decided to investigate cone snails and their venom. Since Dr. Olivera had experience in protein purification, he tried to purify the toxin in cone snail venom to inject it into mice. This was a relatively straightforward project that required no sophisticated equipment or supplies, as opposed to molecular biology, which can get quite complicated and therefore pricey. His original intent was to try to come up with an antidote against *Conus geographus* venom. Later moving to the University of Utah, he had no idea of the important role he would play in the field of venom science. Early on, Dr. Olivera's research group discovered two main components in *C. geographus* venom that accounted for its lethal paralytic properties. One of the components mimicked the actions of a neuromuscular toxin produced by various types of snakes, including cobras. The other had a mechanism of action

---

why the striking biology of cone snails was even discovered was not because these organisms caught the attention of scientists. Rather, it was their looks and appeal as collector's items.

* Always a good student, Dr. Olivera earned an undergraduate degree in chemistry from the University of the Philippines in 1960, a PhD in biochemistry from the California Institute of Technology, and a postdoctoral fellowship at Stanford University. His first research love was molecular biology, at which he excelled. He made quite a few discoveries in the field, including the first description of certain enzymes involved in DNA replication. To this day, he proudly lists his DNA research when listing his academic achievements. I guess you never forget your true first love.

rather similar to tetrodotoxin, which you may remember from an earlier chapter. These were important discoveries, but this research was just the beginning. Dr. Olivera assumed at the time that the cone snail venom he was dealing with contained just a few toxin types. Boy, was he in for a surprise!

By his own account, several of the initial conotoxin-related discoveries by the Olivera lab were due in great part to the ingenuity of his undergraduate students. His group eventually discovered that cone snail venom was composed of multiple factors that could be numbered in the hundreds. More importantly, it rapidly became apparent that these components displayed a wide variety of neurobiological effects, including readily observable behaviors. The ideas of curious undergraduate students gave his research some unexpected twists. Here are a couple of examples.

The first of these talented undergraduates was the late Craig Clark. In Olivera's lab, the standard procedure to test cone snail venoms was to inject mice intraperitoneally (in the gut cavity) and observe any effects. Most of these observed effects were related to neuromuscular block or lethality, in other words, a limited set of responses like paralysis, breathing depression, etc.; nothing very unusual in this milieu. Clark then came up with the idea of injecting the venom directly into the mice's brains. Olivera advised against it, but Clark did it anyway. In my mind, this was the turning point that unraveled the true potential of conotoxins.

When venom was injected directly into mice brains, the team observed a wide variety of effects, many of them unexpected. There was the usual paralysis and death, to be sure. However, when the venom was divided into several components using chemical techniques, and each fraction was injected separately, they were in for a surprise. Some of these venom fractions induced curious behaviors in the mice, such as incessant grooming, scratching, jumping, sleepiness, and what looked very much like behavioral "depression." This depression was characterized by a certain sluggishness or slowness unrelated to paralysis. This is only a partial list of all the different behaviors the team observed. The obvious implication of these results and their initial hypothesis was that each venom fraction affected specific neurotransmitter systems. Further, in very young mice, some venom fractions induced "depression" while the same venom fraction in adult mice induced hyperactivity. This is rather similar to the differential effect of some types of medications in human adults and children.*

One of the challenges that hampered efforts to collect enough conus venom to study was that their original method, namely dissecting snails and squeezing the liquid from the venom sack, did not yield enough material to do more than one or two experiments at the most; these are not big animals, after all. Dr. Olivera wanted to develop an efficient method of "milking" venom from snails. The problem with this idea is self-evident; these guys can kill. Therefore, to grab a live one and essentially squeeze it to milk it is a very dangerous proposition at best. In the Olivera group, they had tried for some years to develop a reliable method when yet another undergraduate, Chris Hopkins, came up with yet another brilliant idea.** A mere retelling of what happened will not do justice to the story. Therefore, let's read the words of Dr. Olivera and his longtime collaborator, Dr. Lourdes Cruz, who relate how they learned to milk snails:

> *"[Hopkins] seemed completely unfazed when we suggested the snail milking project to him. He said he wanted to think about it for a few days, came back and asked for a little cash. He returned with a box of condoms*

---

* It has been long recognized that children are not miniature adults. They have certain specificities in their anatomical, physiological, and biochemical characteristics, due to development. There are a series of medications, for example some barbiturates, that in adults induce their usual calming effects, which in children will induce hyperactivation of the nervous system. This is a highly fascinating topic in itself!

** As a brief aside, I must tell you that you just have to love undergraduate researchers. It is a pleasure working with them because of their enthusiasm and great ideas. I know that from personal experience.

and started blowing one up. He rubbed the inflated condom against a
goldfish and then lowered it into a tank filled with hungry Conus striatus
buried under the aquarium sand. Several came up from the sand, and
one harpooned the condom with such force that Chris let go in surprise.
The condom floated up with the snail still attached to it through its har-
poon and proboscis. The sight of an inflated condom floating at the surface,
with a tethered snail swinging like a pendulum below it, was one of those
moments that should have been recorded with a camera."[131]

Eventually, variations of this milking method were further developed and optimized. Currently, the most common venom collecting
method is to place a small piece of fish inside a plastic tube and cover
it with a membrane of some sort. The snail will strike the membrane and the venom is then injected and collected in the tube. Using
adaptations of this method, scientists are able to milk snails multiple
times so they will have enough material to work with.

Nowadays, in addition to Dr. Olivera's group, there are quite a
few research groups worldwide that are doing phenomenal work in
the cone snail field. In a very real sense, the best is yet to come.[132]
There are some excellent reviews that cover cone snail toxins and
their possible applications in some detail—you owe it to yourself to
seek them out![133]

Cone snails use one of two primary approaches when fishing, the
aptly named *hook and line* and *net* techniques. In the hook and line
approach, the snail burrows under sand or fine gravel and extends
a *proboscis* from inside its mouth. This proboscis is a thin tube that
for all intents and purposes looks like a worm. A fish is attracted
to it, and upon trying to catch the false worm, the snail fires a hollow tooth and through it injects the now quite surprised fish in the
mouth with lethal venom. This happens at ultra-high speeds, in the
range of about 250 milliseconds or so. In the hook and line strategy,
the fish is usually completely paralyzed within a few seconds at most.
At this point the snail dramatically extends the diameter of its proboscis, reels in the fish—just like line fishing—and ingests it. Later
on, the snail regurgitates the hard parts of its meal, like bones and

scales. In a variation of this fishing approach, the snail, most likely by smell, is able to detect a nearby fish and, without being detected itself, stealthily maneuvers its proboscis until it touches the fish's side, stinging it with the same results as above. These strategies comprise two of the main aspects, namely, stealth and venom, that allow cone snails to be active hunters.

The second fishing strategy, the net technique, is even more striking—and quite clever, I might add. A big snail first approaches a small school of fish, extends its proboscis like a net by expanding its diameter, in much the same manner as a snake unhinges its jaw and opens its mouth wide in order to swallow big prey. In this way, the snail can catch more than one fish, and then proceeds to sting each fish individually, followed by eating them. It is worth noting that as the "net" gets closer to the fish, they show no escape activity, and in fact they don't seem to notice the net at all. This suggests that there is more to this strategy than what is immediately apparent. The net approach is particularly interesting because of the startle reflex that fish would display under normal conditions. In other words, if a fish sees what it perceives as a big bag coming towards it, it will undoubtedly swim away from it as fast as it can. A school of fish will normally be gone in less than a second, and yet, that does not seem to happen when being hunted by cone snails. A few years ago, scientists hypothesized that some cone snails that employ this specific hunting method secreted a substance that would not act exactly as a toxin, but rather serve as a mild anesthetic or calming agent that prevented the startle response. This was correct. In 2015, a paper[134] described precisely how at least two cone snail species are able to do this. Researchers were surprised to find that one of the components in the venom of these two cone snail species was insulin, a well-known hormone that regulates sugar metabolism. The interesting detail was that the insulin present in the snails' venom was more like *fish* insulin than the normal snail insulin that these invertebrates use in their own metabolism. Apparently, the snail releases some venom containing a certain amount of fishlike insulin, which then induces a kind of hypoglycemic shock in the fish, rendering them sluggish and unresponsive to most stimuli. The snail can then pick up the fish at its leisure.[135] Pretty crafty, don't you think?

DEADLY BEAUTIES

Probably the first evidence of human fascination with cone snails is a necklace made of cone snail shells discovered in a five-thousand-year-old Mesopotamian tomb.[136] Because of their unusual beauty, these shells have always been in high demand by enthusiastic collectors. For example, in 1796, a rare cone snail shell of the species *Conus gloriamaris* fetched a higher price in a public auction than a painting of the posthumously famous artist Johannes Vermeer. Cone snails are often portrayed in drawings and paintings. Two more recent examples show the influence of cone snails on our imagination: a depiction of one cone snail species (*Conus textile*) as the murder weapon in an episode of the 1970s television series *Hawaii Five-O* titled "Cloth of Gold"[137] and more recently, in 2010, a generic cone snail venom as a murder weapon in the *CSI: Miami* episode "Sleepless in Miami."[138] Police shows are nice and all, but I must admit a strong personal predilection for science fiction, and cone snails have also found their way there. The venom of a cone snail species (*Conus purpurascens*) had a small cameo in the 1997 *Jurassic Park* movie sequel, *Jurassic Park: The Lost World*, where the venom was a component injected by darts,[139] supposedly allowing the characters to kill dinosaurs.* These TV shows and movie examples are

---

* Of course, no one has ever performed a conotoxin toxicity bioassay with a dinosaur. Nonetheless, this actually elicits a great memory of one of my professors at grad school. In one of my courses, Prof. Ron Harris-Warrick, at the time the chair of Neurobiology & Behavior at Cornell University, told us about this specific conotoxin and its connection to the *Jurassic Park 2* movie. He even treated us by acting out the scene! Good times . . .

good examples of how scientific discoveries find their way into our general culture.

While there were no cone snails in the dinosaur-ruled world (which, as we know, ended well before these snails came into existence), cone snails predate and of course have coexisted with humans, and people tend to like to pick them up. Unfortunately, in some instances the snail stings, sometimes with fatal consequences. The first written record of a cone snail–caused death comes from the German/Dutch botanist Georg Eberhard Rumphius in 1705. He witnessed the death of a woman from the island of Banda, near Indonesia, who was stung upon picking up a cone snail from the beach. In his own words:

"A native woman kept this shell in her hand after she had picked it up when hauling in the seine at sea. As she went to the beach she felt a faint tickling sensation in the hand which crept slowly through her entire body—she died on the spot."[140]

After that, there were several reports in scientific literature (mostly from the 1800s and early 1900s, all by physicians) of cone snail stings and fatalities. The earliest attempt to investigate cone snail venom was by Dr. Louis C. D. Hermitte[141] (1894–1961) of Mahé, one of the Seychelles islands east of Africa. In 1946 he reported the 1932 case of a man (a shell collector, of course) who picked up a cone snail from a shallow pool and proceeded to try to clean it of debris by scraping it with a pocketknife. As you would think, he was stung, and almost immediately his arm went numb and within the hour his entire body was numb as well. He experienced visual difficulties and dizziness, among other symptoms. After a few hours he felt somewhat better and was taken to see Dr. Hermitte, who treated the collector, and the man fully recovered over the next few days. The cone snail that stung him was later identified as *Conus geographus*. This gentleman was rather lucky, because *C. geographus* is the cone snail most associated with human fatalities; according to most sources, about 70 percent of untreated patients die from its sting. In a similar case to the one that Dr. Hermitte described, in 1936, another man vacationing in Australia picked up a cone snail, and he too proceeded to clean its shell with a knife. He was stung, but was not as lucky as Dr. Hermitte's patient, since he was dead in five hours.

Dr. Alan J. Kohn* is the undisputed grand master of cone snail biology as well as of their *systematics* and *taxonomy* (ornate terms to describe how organisms are biologically classified). Dr. Kohn is an invertebrate biologist and currently a professor emeritus at the University of Washington. He became interested in cone snails in the 1950s, and he and his collaborators were the first to record and publish a description (including pictures) of one cone snail species (*Conus striatus*) stinging, catching, and eating a fish. They observed for the first time the proboscis lure, the actual sting and paralysis of the prey, and its subsequent ingestion (as well as the beginning of digestion!). Dr. Kohn also worked on the initial characterization of cone snail venom and has published numerous articles and books on cone snail biology. Sadly, he is almost never mentioned when the conotoxin story is told, but to this day, he is still active in this field of research.[142] Prior to his work, the only hint of fish-eating snails was the discovery of fish remains in dissected conus specimens found in 1943.

Research in this area continued with a series of experiments by the late Dr. Robert Endean, of the University of Queensland in Australia, and his collaborators. Dr. Endean led what probably was the first systematic study of cone snail venoms. In a series of studies in the 1960s, he and his colleagues conducted a survey of about fourteen different cone snail species to "map" their prey specificity. Based on that research, they were able to get an idea of which species of cone snails were likely to be dangerous to humans. Dr. Endean's

---

*Yes, his last name kind of sounds like "cone" if pronounced a certain way and one could think of it as a mere coincidence, of course. However, interestingly, genealogically the origin of the "Kohn" last name is actually "Cone." I love these silly coincidences!

group also pioneered the idea that these venoms had different active components as well as illuminated many other pieces of information that provided important leads for other researchers.[143] I am confident that cone snails still have many secrets to share with us.

» SPITTING SPIDERS

Nightmares are the worst. They come in many nasty flavors and seem to be a reflection of our most intimate fears. Some nightmares sadly mirror our daily lives and can recall the loss of a loved one or a similar traumatic experience. Still other nightmares reflect our primeval fears. Although we did not remain as such for long, our history as a species began with us as prey. Individual humans devoid of technology are still prey in the relentless circle of life.

A common nightmare related to our past as prey is to feel trapped. This one comes in many varieties, but to me, one of the worst involves being trapped in a spiderweb or similar sticky thing. It is an all-too-imaginable feeling, the inability to move, the dread of the predator that is certainly coming, the desperation of trying to escape. In such a nightmare, you may be able to avoid your fate if you steer clear of the web, but what if a predator were capable of shooting a sticky web right at you? Some two hundred of the forty thousand or so spider species described are able to do just that. Virtually all species that display this behavior belong to the *Scytodes* genus and are very descriptively called the "spitting spiders." These spiders were originally classified in the 1800s, and scientists have known of their silk-spitting capabilities since the 1920s or so. It is only recently that we have been able to elucidate some further aspects of their interesting behaviors. These spiders are—thankfully!—small; their bodies are about a fifth of an inch at the most, and they are very common.

The general behavior of spitting spiders consists of projecting a sticky web onto their prey* in an intricate zigzag pattern, which

---

* Yes, just like Spider-Man.

optimizes their ability to target their victim. They are able to project their silk at high speeds, typically at about ninety feet per second! The prey is then rapidly covered and immobilized. However, there's more: As the liquid silk is ejected from the spider's body, it solidifies when exposed to air, causing its fibers to be shortened by about half their length. Hence, the prey is trapped by the glue on the silk, and when the fibers shrink, they tighten and hold the prey in place.* Science-fiction movies have portrayed such captures quite graphically! Once the prey is immobilized, the spider proceeds to inject the venom by biting the victim, killing it, and at the same time initiating digestion. This is pretty much the standard for how web-using spiders prey on their victims. Amazingly, some spitting spiders add "injury to the insult" (yes, I know, I reversed the saying) after trapping their prey with their spit-sticky silk. In addition to the usual components intended to immobilize the prey, their silk also contains venom, which is what actually kills the victim.

It is interesting to compare how the relative strategies of diverse species of spiders have evolved. Consider the passive way in which a web-building spider waits until a victim is caught in its silken trap, giving the spider essentially all the time in the world to get to it and kill it. A spitting spider seems impatient in comparison; one could almost say that spitting spiders are more proactive in their approach. Both strategies display advantages and disadvantages depending on the prey in question, but it is evident that they are both successful strategies; after all, both types of spiders are successful survivors, or they wouldn't be here today. One thing is for sure: we have barely scratched the surface of these intriguing animals.[144] For another

---

* Again, science fiction seems to imitate nature. In the 2004 movie *Alien vs. Predator*, the predators' weapon complement includes a net that shrinks upon trapping the prey, helping to immobilize and kill the hapless victim. Moreover, just like the spitting spider's net, the predators' net seems to have some chemical (or irradiates some energy, it is not clear) that burns the victim.

delightful nightmare, keep reading. Next up is a fascinating animal whose paleontological origins imply hallucination!

» HALLUCIGENIA AND THEIR DESCENDANTS: VELVET WORMS

The Burgess Shale is one of the most iconic paleontological sites worldwide. It is located in the rocky mountains of British Columbia, quite isolated from the rest of the world. There are other sites from the same geological era and thus containing similar fossils, most of them hailing from about half a billion years ago. These sites are collectively called "Burgess Shale–type" sites. The significance of these fossil deposits to natural history is difficult to overstate—in them we'll find some of the earliest paleontological evidence for a biological event that puzzled scientists for the longest time, and is still a source of academic discussion: the so-called Cambrian Explosion.[145] This episode in the history of biology on our planet concerns the sudden—in geological terms, that is, in this case over ten million years or so—appearance of a wide variety of organisms with diverse structural features and body types. Some of these organisms, like crustaceans, are still with us. However, there were quite a few frankly weird animals that have no modern relatives at all, as far as we can tell. As a matter of fact, many of the animals found at the Burgess Shale sites do not seem to have left any modern descendants, a poignant reminder of the sometimes ephemeral nature of biological life.

The credit for the 1909 discovery of the Burgess Shale and its wonderful fossils goes to Dr. Charles Doolittle Walcott.[146] He was an honest-to-God "Dr. Doolittle" who, even though he did not talk directly with the animals that he discovered (they were, after all, long dead), was nonetheless able to investigate them in an attempt to unravel the mysteries of early complex life. He was very successful at this. Over the years, other scientists have refined his insights and reinterpreted his findings, but that does not diminish his accomplishments in the least. One of the animals that he found in Burgess Shale deposits is a wonderfully weird creature appropriately called *Hallucigenia*.*

Hallucigenia were small—up to about one and a half inches long—wormlike organisms, with some features that you would not expect to see in a worm. For example, the most common type of *Hallucigenia*-like organism had seven pairs of spines radiating from a central "tube" on one side and a series of tentacles on the opposite side. The original 1977 reconstruction by the English paleontologist Simon Conway Morris had *Hallucigenia* standing on its spines. This interpretation turned out to be incorrect. He essentially displayed the organism upside down (don't hold it against him; I would have likely portrayed it upside down as well). The mistake was cleared up a bit in the early 1990s with the discovery of better preserved *Hallucigenia* fossils by Lars Ramskold and Hou Xianguang, and culminated in 2015 with the work of Martin R. Smith and Jean-Bernard Caron. After them, it was pretty evident which part of *Hallucigenia* was up, which part was down, which end was the head, and which end was the tail.**

Once these researchers clarified the correct spatial orientation of the animal, it was even more apparent that they were dealing with a rather unorthodox type of worm. At the time, the general consensus was that *Hallucigenia* did not have any modern relatives. There were some ideas about a possible taxonomical relationship between *Hallucigenia* and two living organisms: the velvet worms and tardigrades. This link was proposed in the early 1990s based on molecular biology studies.[147] But there was no other evidence of this until a paper published in 2014 provided hard morphological proof of this relationship.[148] In this paper, the authors studied in detail the anatomy of the

---

* On account of its more-than-weird appearance, which is reminiscent of an animal that you would see in a dream but would not exist in reality.

** The story of how scientists found which end of *Hallucigenia* was "up" is summarized in a very nice article: www.theverge.com/2015/6/24/8838169/hallucigenia-worm-fossil-nature-study-2015.

claws on *Hallucigenia*'s legs. They found that the specific morphology of these claws was very similar to the claws and jaws of certain modern organisms, the *onychophorans* (this term means "the clawed ones"), most commonly known as velvet worms, with *peripatus* as the best-known example. Based on further phylogenetic analyses, the authors were also able to determine that another type of modern organisms, the tardigrades or "water bears," seemed to be related to *Hallucigenia* as well.

Water bears are amazing in their own right. These microscopic tough guys are resistant to radiation and chemicals and can go temporarily dormant when life becomes difficult—just to underscore how resilient these guys are, note that some specimens of this animal were taken into orbit and exposed to the hard vacuum of space, unshielded solar radiation, the works. And many survived the ordeal![149] I do not think that there is any other animal capable of such a feat.

In addition to their distinguished lineage,[150] velvet worms are unusual and fascinating organisms. They look like elongated tubes with a pair of antennae on one end and come in a range of sizes, from a couple of millimeters to individuals close to 8 inches long. There are fossil examples preserved in amber dating from about one hundred million years ago, but these worms are likely much more ancient than that. One of the reasons we believe this is because their distribution is so wide; they are found pretty much everywhere there is some kind of tropical climate. This fact has led taxonomists to believe that velvet worms date from the time when all continents were joined in one supercontinent (Gondwana), about five hundred million years ago. There is an even more ancient example of a peripatus-like animal that was found at the Burgess Shale by Walcott himself. He named it *Aysheaia*, and it is one of several similar animals found in the Burgess Shale.

Peripatus come in a variety of different colors,[151] but my favorite is an azure-looking one, *Peripatoides indigo*. It is absolutely stunning![152] They do not move like worms at all, rather, they actually walk using almost comical-looking legs along both sides of their bodies. Speaking of their legs, they have a rather unusual characteristic: excretory organs function as kidneys in each of their legs, with openings to the external environment. Yes, you could say that most velvet worm species pee through their legs. They also display many unique characteristics related to their nervous system and their general physiology.[153] They are not very long-lived—their life span is two to three years—but most species display some version of live birth, producing litters of up to forty tiny worms. Surprisingly, some species are known to care for their young and live in social colonies.[154]

All known species of velvet worms are predators, yet they have no eyes at all. They seem to track prey by smell, and they are ambush predators! When a peripatus gets within range of possible prey, it rears its head back and sprays the unfortunate victim with a sticky slime* that immobilizes it. The peripatus's hunting technique is shared by no other organism on Earth.** Once the prey is trapped, the velvet worm devours it using its sickle-like jaws, a distinct characteristic of these animals. Moreover, as it eats the prey—slime and all—instead of digesting the slime, it recycles most of it. This makes sense, because as a general rule such specialized secretions are metabolically expensive.[155] You are very welcome for that information. Sweet dreams!

---

* Yes, again, just like Spider-Man.

** Although I have witnessed a similar strategy used by planarians, my favorite invertebrate. They actually secrete a sticky slime as well that traps small prey like water fleas. Then the planarian proceeds to coil itself around the small crustacean and eats it. Again, nobody told me of this behavior and I didn't read about it in the scientific literature. *I have seen it!*

CHAPTER 6

THE VERY BEST
SURVIVAL TACTIC
OF THEM ALL

B elieve it or not, evolution is not always about competition. The *E* word is famously linked with competition, from Herbert Spencer's well-known phrase *"survival of the fittest"** all the way to *"Nature, red in tooth and claw,"* immortalized in Alfred, Lord Tennyson's poignant poem "In Memoriam A.H.H.," written in honor of a lost friend. Tennyson's poem tone was strongly influenced by the evolutionary thinking of the mid-1850s,[156] thinking that culminated in the publication of Darwin's *Origin*, his 1859 magnum opus, a book that has been equally praised and despised over the years, nonetheless indubitably changing the face of biology forever. At face value, it is perfectly understandable that evolution

---

* This phrase is oftentimes—yet incorrectly—attributed to Charles Darwin. In actuality, the phrase was the brainchild of Darwin's contemporary, Herbert Spencer, who was also a natural philosopher and who coined the phrase in 1864. Come on; Darwin couldn't have possibly thought of *everything*!

145

is frequently seen as driven by competition; this is an all-too-evident aspect of life on Earth. No question about it, competition is an important engine of evolutionary change. Natural selection weeds the ill-adapted branches of the biodiversity tree,* and there's no denying that competition is at least one of the processes that drives evolutionary change. In the natural world, competition often means death for one of the competitors; as I said earlier in this book, death gives directionality to the evolutionary process. In other words, pruning is oftentimes useful to make the proverbial tree stronger.

On the surface, it might seem that cooperation and even sacrifice of a number of organisms in order to benefit others is evolutionarily counterintuitive, since the "cooperator" or "altruistic" organism will see a diminished opportunity of passing on its genes. However, when scientists examine the world of life in some detail, it is acutely evident that cooperative interactions are present at every single level of biological organization, all the way from molecules to behavior. Solid theoretical models predict—correctly—that cooperation is at least as effective as competition as far as survival is concerned. And there are at least five general principles that are thought to govern the evolutionary success of cooperation: kin selection, direct reciprocity, indirect reciprocity, network reciprocity, and group selection. A thorough explanation of these mechanisms is beyond the purview of this book,** but I will try to give you a taste of how cooperation benefits living beings by using examples from a wide variety of organisms.[157]

There is plenty of evidence to support the idea of cooperation—within and between species—as a key tool for survival, given that associations between at least two distinct species with interactions that look very much like cooperation are found everywhere in nature,

---

* Under a specific set of environmental influences, that is. If those influences change, a prior disadvantage can easily turn into a distinct advantage. This is all in the context of genetics and the environment, or as it is most commonly known: *nature and nurture.*

** And there are quite a few (very) technical and mathematical books exploring the contribution of each of these aspects to the evolutionary success of species.

which would not be the case were cooperation not a useful feature. These associations are referred to as examples of *symbiosis*,[158] a term that essentially means "living together," and it is in fact observed in every kind of environment examined so far: terrestrial, marine, and everything in between. Over time, scientists have discovered that symbiotic relationships have played a central role in how many species appeared on Earth, directly influencing biological diversity. In turn, this diversity displays the kind of variation that might include previously unknown (to us) survival mechanisms, as well as the development of survival mechanisms never before seen in the history of nature.

Symbiosis can appear in several incarnations. One of them is *mutualism*, which occurs when both organisms benefit from the symbiotic relationship. An alternative occurs when one of the partners tends to be harmed and sometimes killed by the relationship. This modality is very well-known and is called *parasitism*. Finally, there is a kind of "intermediate" relationship between organisms called *commensalism*, in which one of the participating organisms benefits and the other is neither benefited nor harmed; it is just there. There are a few other cooperation modalities with their own subtle particularities, but the point that I want to get across is that association between organisms is hardly unusual in nature. Moreover, we could argue that symbiosis is one of the most fundamental biological principles that we have discovered so far. Let's see some examples of this phenomenon.

As a nature lover, you may already know about the close relationship between some hermit crabs and certain species of sea anemones. The sea anemones get a free ride and some food scraps left from when the crab feeds, along with other miscellaneous benefits. On the other hand, the hermit crab literally carries around a bodyguard wielding poisonous darts. Pretty good arrangement, don't you think? Or consider the "agreement" between Nile crocodiles and Egyptian plovers (plovers are the cute little birds that pick the crocodiles' teeth). You'd think that only animals with a death wish would willingly get within striking distance of a crocodile's mouth, yet in this case, the birds get a free meal and the crocodiles get some dental hygiene. There are even certain types of small marine fish—collectively called "cleaners"—that routinely work at "cleaning stations" that much bigger fish visit specifically for the attentions of the cleaners. The cleaners safely pick and eat parasites and debris off the skin of the bigger fish, and no, as a rule, the small fish do *not* get eaten. Furthermore, some cleaners do not have proper stations but instead follow the bigger fish, acting as a private cleaning crew. Even some of the most feared predators of the sea, like barracudas and sharks, including great whites (which are probably *the* prototypic marine predator), associate themselves with much smaller fish in a nonpredatory way. In the case of the great white, oftentimes the smaller fish are pilot fish (*Naucrates ductor*) that follow the shark and do for it exactly what the plover birds do for Nile crocodiles. As you may imagine, the type of relationship seen between anemones and crabs, plovers and crocodiles, and cleaners and sharks—as well as many other similar interactions described in nature—is an example of mutualism.

Mutualistic cleaning behavior is not limited to vertebrates. Certain invertebrates, including several species of shrimp, also engage in fish cleaning. Surprisingly, the fish that benefit from this relationship even allow the cleaning shrimp to pinch their skin, sometimes even making small cuts in their efforts to pluck away parasites. There is even a species of shrimp that specializes in cleaning another fearsome fish, the moray eel. Curiously, it seems to me that this specific example of mutualism must have developed relatively recently, since eels sometimes shake their heads in a rather "emphatic" way, as if they were annoyed by the shrimp's plucking behavior. Also, although the eels generally tolerate the shrimp, a significant number of shrimp are frequently found in eels' stomachs. Again, either this is a "newer" relationship in the evolutionary sense or else the moray eels are really, really mean fish.

To illustrate how cooperation has the capacity to dramatically enhance the probability of survival of a species, we have to go no further than ourselves. Nobody denies that of all the creatures on this wonderful planet we are in no way the biggest, the baddest, the fastest, or any other "-est" that we can think of, with one true exception that I'll address in a moment. Before presenting that exception, I would like to point out that despite the relative lack of physical advantages that we humans display in the biological sense, we have spread all over the planet, dramatically changing it in a variety of ways. These changes can be interpreted as positive or negative depending on the specific context, but that is a story for some other time.

As you are probably already thinking, our intellectual and reasoning capabilities are what set us apart from every other organism on Earth. We are "the smartest of them all," at the very least in the technological sense. Humans evolved a knack for inventing things—and that is what has largely made a difference. In other words, the secret of our survival success can be safely attributed to our intellect; there is no denying that. Again, as far as we know, our "out of scale" mental capacities set us apart from everything else in nature, but make no mistake, our mental prowess amounts to pretty much next to nothing in the absence of cooperation between members of our species. For example, thousands of years ago, a lone man had virtually no chance of surviving when confronted by, say, a saber-toothed tiger, provided that both man and beast fought using just the physical weapons that nature gave them. The invention of technology in the form of spears, or even rock tools, may have helped a bit, but the odds were invariably better for the human if he had a few buddies as backup.

Another example of how cooperation is what "did the trick," as it were, for humans, concerns the bright idea a person dreamed up of steering a group of mastodons or mammoths over a cliff in order to kill them and subsequently have them as dinner. Again, a lone person could almost certainly not have done this, since to a herd of mastodons, a single human would be next to nothing. It stands to reason that a solitary guy could be—and very likely oftentimes was—casually brushed aside by the mere sweeping of a trunk, or worse, might be deliberately stomped on.* Only through coopera-tion with his—her?—peers could the idea finally bear fruit. History tells us that strategies like this did work—only too well, I may add. Eventually, our combined efforts led to the development of weapons and other technologies, contraptions that over time helped secure our dominance over virtually every single ecosystem of this planet for good, bad, or worse. We are not limited by geography either, and this is more than a function of transportation technology. History is full of examples indicating that the presence of a big, deep ocean or a high, massive mountain range is simply not a deterrent to most of us. Quite the contrary, our story as a species is, in big part, the story of wanderers who crossed the oceans or climbed the mountains simply "because they were there" or "to see what was on the other side."

In a real sense, cooperation is the most sophisticated survival mechanism of them all. I have tried to illustrate how humans have benefitted from our cooperative tendencies, but as we saw above, we did not invent cooperation: other animals beat us to it. Furthermore, even these other animals were not the first to engage in cooperative behaviors. The story of cooperation started with single, simple cells.

THE FIRST CASE OF
STOCKHOLM SYNDROME**

In a literal sense, living beings have been cooperating with each other for the last two billion years or so; to give you some perspective, this is almost half the age of our planet! In Chapter 2, we talked about

---

* And in the immortal line of the late Bill Paxton in the 1986 movie *Aliens*: "Game over, man; game over!"

** Stockholm Syndrome is a psychological phenomenon described as early as 1973 that delineates the pathological development by kidnapping victims of empathy and even affection toward their captors. In extreme cases, the victims actively defend the behavior and motives of those who kidnapped them. I am, of course, using it as a metaphor here.

prokaryotes and eukaryotes. You may recall that prokaryotes are uni-cellular organisms that include ancient living beings like bacteria and archaea. On the other hand, eukaryotes are more intricate organisms, most of them multicellular, with cells that display higher complexity by virtue of an ancient event that forced the hand of some of the prokaryotes that used to live here billions of years ago. Eukaryotic cells can be considered as "cells within cells" in the sense that they all contain a particular organelle, the mitochondrion, which as you may recall from your biology classes is the organelle that is the main source of our chemical energy in the form of ATP.* Nowadays, eukaryotic cells cannot live without mitochondria and vice versa. This is a fact of life that is explained by the *endosymbiont theory*. This particular example of the cellular symbiosis that characterizes the origin of eukaryotic cells was described in the scientific literature very early in the history of biology, in the 1800s. However, the idea of cellular symbiosis was not taken seriously at all by the general scientific community until about the 1960s. And then it was only brought to the attention of modern science due to the pretty much single-handed efforts of one of the most underrated scientists of our era: the late Dr. Lynn Margulis. There is no space in this book to properly describe her scientific contributions, but suffice it to say that, as has been the experience of many "game changers," when she presented her ideas to the general scientific community, she was respected and even liked by a few, she was ignored by most, and she was ridiculed by more than a few, only to be proven right in the end. Endosymbiont theory by itself was controversial, but it is undeniable that at least part of the resistance to her arguments was simply because she was a woman.[159]

It is important to point out that the specific steps that led to the original endosymbiosis event (or events, as this may have happened more than once; alas, only one of such events seems to have "taken")

---

* Photosynthesizing plants have yet another organelle—the chloroplast—that became associated with certain early eukaryotic cells. This is how green plants came to be.

on Earth are still unknown, and are perhaps forever unknowable, but we can speculate on some of the basic aspects of what occurred. Not surprisingly, one of the best popular exponents and explainers of this biological event was, of course, Dr. Margulis herself.[160] This is a complicated topic, but I can summarize some of the main points for brevity's sake. The story seems to have taken place a couple of billion years ago, when a biggish prokaryotic organism ate a bunch of smaller prokaryotes. We know for sure that one of these prokaryotes was a bacterium, and the other one was an archaeon, but we just do not know which was bigger.* Nonetheless, on any normal day, the smaller guys would be rapidly digested, or alternatively, the smaller guys would infect the bigger guy, killing it. There was no second place—or even a participation trophy—one of the adversaries would inevitably die. However, it seems that one fine day, after the ingestion (or the infection) event, probably one or perhaps even both types of cells "refused" to die and, furthermore, the smaller cells established a permanent "base camp" inside the bigger organism. This meant that eventually, over several generations of both the bigger and smaller classes of cells, the two became interdependent. One could not live without the other. Was this perhaps the first—and quite an extreme—example of Stockholm Syndrome? One thing is true: the series of events that led to the development of eukaryotes most likely paved the way for multicellularity.

The development of true multicellularity extended the possible ranges of size in living organisms. No longer did living beings have to be limited by the relatively short range of the diffusion of chemicals like oxygen. They went from clusters of pretty much identical cells to ensembles of different cell types that specialized in distinct

---

* However, genomic research seems to indicate that the closest modern relatives of all mitochondria are certain types of bacteria, generally belonging to the pathogenic bacteria *Rickettsia*. These bacteria are responsible for causing typhus and Rocky Mountain spotted fever among other infections. It seems that bacteria, as opposed to the archaea, were likely the little guys after all!

functions. This was followed by the organization of tissues, organs, and eventually whole organisms with mutually interdependent parts. In essence, endosymbiosis made organisms like us possible, in our case with all thirty-seven or so trillion of our cells. Just imagine how many cells are in a blue whale!* One may be excused for thinking that endosymbiosis is a quite speculative idea based on the extreme improbability of such an event, added to the fact that if this is the way it happened, it happened so long ago that any clues about it must have been well-buried by the sands of time. In other words, this could be the ultimate cold case. However, a series of important experiments that began in the 1960s and went on for at least thirty years proved without a doubt that endosymbiosis is not only possible, but it can also take place in a time period that is significantly shorter than a proverbial geological eyeblink.

In 1966, the microbiologist Dr. Kwang Jeon, now a professor emeritus at the Department of Zoology at the University of Tennessee, suffered what is a not uncommon, yet always undesirable, occurrence in a microbiology laboratory.[161] His cell cultures became contaminated with other organisms. This meant that the cells that he used in his research were infected by other cells, the nature of which I will explain in a moment. To a microbiologist, the unintended contamination of his or her cells is never a good thing, and yet for Dr. Jeon, it was serendipitous. Dr. Jeon's experimental organisms were amoebas (*Amoeba proteus*), a eukaryotic organism that happens to be unicellular (there are a few of those). Amoebas are the proverbial flag bearer of the biology laboratory, the quintessential living being that all of us biology enthusiasts are likely to see when looking down a

---

* Or, if you don't feel like imagining, I can give you a very rough idea: A typical man weighs about 200 pounds. A fully grown blue whale weighs about 200 tons (which translates to about 400,000 pounds). Thus, a blue whale is roughly equivalent to 2,000 people. If a typical human has 37,000,000,000,000 (37 trillion) cells, a typical blue whale has about 37,000,000,000,000 x 2,000 = 74,000,000,000,000,000 cells (a whole lot).

microscope for the first time. By the way, no one ever forgets the first microbe they see through a microscope. Mine were *Euglena*.[162]

One day, Dr. Jeon looked at one of his amoeba culture samples under a microscope, and it was immediately apparent that there was something amiss. A researcher who uses cells in his or her research learns very quickly to recognize when the cells are healthy and when they are not. Dr. Jeon immediately realized that there was some kind of infection, tipped off by the fact that there were a high number of dark dots within each amoeba, filling about 35 percent of each cell's volume (this translated to about 150,000 dots; I did not find out exactly how he counted them). Upon closer examination it became evident to him that each dot was an individual bacterium. This happened to be a previously unknown strain of bacteria, and Dr. Jeon dutifully called them the X-bacteria. In most circumstances, amoebas have nothing to fear from bacteria. In fact, amoebas *eat* bacteria. However, it seems that, in this particular case, the bacterial strain was so aggressive that it invaded the amoebas in unmanageable numbers. We can think about this from the perspective of an anteater (yes, an anteater; bear with me for a moment). An adult anteater can easily handle an ant colony of a few thousand individuals, no problem! However, a particularly unlucky anteater that stumbles upon a hypothetical ant colony of say, one billion—particularly aggressive—fire ants*... let's just say that the odds will not be in the anteater's favor.

Back to our amoebas. Under virtually any other circumstance, such an aggressive bacterial infection would wipe out the amoeba population entirely. It had happened before in many laboratories and would likely happen again; usually the end result of these infections is that you lose your cells and have to start your experiments all over again. In this particular instance though, not all amoeba perished; some indeed survived! This fact did not escape Dr. Jeon's powers of

---

* By the way, no ant colony is that big; the record seems to be about three hundred million individuals or so; still, a very respectable number!

observation, and he took samples of the surviving amoebas and cultured them. Eventually, he was able to raise a new amoeba strain that in turn learned to cope with invading bacteria—though a smaller number (about forty-two thousand, roughly occupying 10 percent of the cell's volume).

Over time, the alert Dr. Jeon and his collaborators noticed something quite unexpected about the new amoeba strain: they did not only endure having X-bacteria living within them—they now could not live without them! Upon careful study, the team was able to prove that these amoebas had developed into a strain that could not survive without the invading X-bacteria; if you took the bacteria out of the picture, the amoebas died. Dr. Jeon and his collaborators demonstrated this interdependence using microsurgical, biochemical, and genetic methods. After exhausting virtually all other possibilities, they were left with an inescapable conclusion: they had witnessed a modern example of endosymbiosis! One amazing detail of this story is that the new bacteria-friendly amoeba strain developed over a mere eighteen months, not even a hundredth of an eyeblink, geologically speaking.

It was remarkable that Dr. Jeon and his colleagues happened to catch a glimpse of evolution in action. Dr. Jeon's body of work does not prove that the endosymbiont theory exactly describes how eukaryotes came into existence on Earth, but it does demonstrate the plausibility of this mechanism.* In scientific parlance, he provided *proof of principle*. The reasoning goes thusly: if a kind of endosymbiosis did happen in a mere few months in Dr. Jeon's laboratory, what could happen in a couple of billion years on a planet as big as Earth? At the very least, this is food for thought.[163]

---

* Incidentally, how Dr. Jeon's research has not earned a Nobel Prize yet I could never understand. To be fair, I do not think that there is a suitable category that fits this particular discovery.

NO BRAINS, EARLY SOCIETIES

It would be perfectly understandable to be blown away by the thought of how improbable was the endosymbiotic event that led to multicellular organisms and eventually to us. Indeed, this fact invites us to ponder our existence at many levels, from the biological to the purely philosophical. And as improbable an event as it was, the amoeba and X-bacteria story shows that such events repeat across biology with unexpected frequency. For example, there are many other interesting cases of endosymbiosis that include the interactions of bacteria and algae with organisms as varied as protozoans and plants.[164] But it is one thing to talk about "cells within cells," and another thing to talk about billions, even trillions of cells cooperating with each other to reach a common goal, and in some organisms forming a distinct entity, which is what I would like to talk about next.

Endosymbiosis may have cleared the way for an extreme class of "intercell" cooperation, namely multicellularity, but it is important to point out that prokaryotes have been cooperating with each other for a very long time, likely well before the appearance of eukaryotes. In other words, some bacteria have social lives, and as fascinating as this fact is, this may not be good news for us.

We knew of social bacteria long before we thought of them as social. For example, early biologists described complex macroscopical structures called *fruiting bodies*. These structures were recognized as bacterial colonies as early as 1892. Fruiting bodies were formally described by the botanist Roland Thaxter, who observed them in a particular bacterial species, *Myxococcus xanthus*, of the general class of myxobacteria—colloquially called "slime bacteria."[165] Later on, in the 1970s, Dr. Martin Dworkin, at the time working at the University of Minnesota, described a very interesting series of behaviors typical to some myxobacteria—behaviors that he correctly interpreted as an early example of cooperation,[166] as these bacteria seemed to be coordinated predators! Dworkin observed that the myxobacterium *Chondromyces crocatus* had the capacity to form large colonies that travel together like a giant amoeba ("giant" in this context meant a couple of millimeters long; bacteria tend to be about a thousand times smaller than that). When *C. crocatus* detects prey, it coordinates a massive simultaneous release of slimy secretions containing digestive enzymes from some of the individual bacterial cells, which digest their target. Think about the cells that line your stomach and release digestive enzymes, acids, etc., helping to digest what you eat. We can think of myxobacteria as having an "external stomach" of sorts.

For more than twenty years after the publication of his observations, Dr. Dworkin's work was considered a biological curiosity, nothing more. Truth be told, most scientists found this behavior quite interesting while at the same time thinking of bacterial sociality as a rare occurrence, a mere exception within the much larger microbial world. However, since the 1990s, it has been increasingly evident that bacterial cooperation is much more common than we thought. There are many newly described examples of cooperation between individuals in bacterial colonies. In fact, so much evidence was accumulated in favor of the renewed interest in this bacterial property that in 2005 Drs. Matthew Parsek and Everett Greenberg coined a new name for this new subdiscipline of microbiology. They called this new science *sociomicrobiology*.[167] This term was of course inspired by the more widely known science *sociobiology*, the brainchild of Harvard University's Edward O. Wilson.[168] In contrast to sociobiology, which explored the association of genes and behavior as a fundamental property of biology, sociomicrobiology was born in response to a practical and very relevant public health problem: we need new antibiotics and new ways to fight bacterial infections,[169] since one of the most urgent crises in biomedicine today is the emerging phenomenon of antibiotic resistance, as discussed early on in this book.

One of the best-known examples of antibiotic resistance is the development of MRSA (methicillin-resistant *Staphylococcus aureus*). *S. aureus* is a common type of bacteria that naturally lives on the skin of many mammalian species, including humans. Of course you are familiar with a common mantra of real estate: "Location, location,

location." The same is true for many members of the bacterial world. When *S. aureus* is where it is supposed to be, there is no problem, but when it penetrates the skin, it can cause infections that range from cellulitis to full-blown pneumonia. Up until very recently, common antibiotics would be effective to treat such infections. However, a bacterial strain resistant to antibiotics can infect a human with impunity, possibly resulting in the death of the person. MRSA is just one of the many examples of modern antibiotic resistance.

In this framework, two particular bacterial social behaviors kindled the interest of scientists and, of course, physicians: the phenomenon of *quorum sensing* and the formation of *biofilms*.[170] The term "quorum sensing" is not even a quarter of a century old at the time of this writing, but in retrospect, scientists recognize the early discoveries about myxobacteria as perfect examples of this phenomenon. What exactly is quorum sensing? In a nutshell, it is the capacity shared by a number of bacterial species to sense population changes as a colony, specifically the number of other bacteria nearby (population density), and then modify their overall behavior according to this information.

Whew, I need breather after that sentence! Perhaps a sports comparison will help clarify the thought. Let's imagine a hockey team. Their ultimate objective is to get the puck into the net. To do so, each player should be aware of all the other players' positions (and velocity).* Only by communicating with each other before and during the game can they maximize their chances of winning. Think of each player as an individual bacterium. Bacteria sense the location of their fellow bacteria through cell-to-cell signaling based on the concentration of specific chemicals in the environment. A variety of substances have been proposed as candidate molecules to induce the quorum sensing response.

---

* Which cannot be achieved at the molecular level (for all my fellow physics enthusiasts). At the subatomic level, you can know the position or the speed of a particle like an electron, but not both. I'm done geeking out.

To discriminate between candidate compounds, scientists developed five criteria that must be fulfilled in order to consider a specific compound as having "quorum sensing–inducing" properties,[171] namely: (1) its synthesis is triggered by a specific environmental change, (2) it accumulates in the immediate neighborhood of the bacterial cells, (3) it binds to specific bacterial receptors, (4) such binding in multiple bacterial cells induces a collective response, and (5) such a collective response must be different from the mere metabolic manipulation of the substance by the bacteria.

One of the main bacterial collective responses induced by such a chemical is the expression of specific genes within the individual bacterial cells, which in turn can change and control how pathogenic (disease-causing) the particular bacteria can be and if they will develop antibiotic resistance, for instance by the development of biofilms. In biofilm formation, bacteria produce molecules that can combine with environmental material, acting as scaffolding on which bacterial colonies can thrive.[172] Biofilms notoriously make bacterial colonies resistant to antibiotics, and they are therefore a furiously studied target for the discovery of new medications against bacterial infections, as many bacterial-borne disease conditions appear to be a direct consequence of their biofilm-producing capabilities.[173] The study of bacterial societies is not limited to the few pathogenic ones, however. Modern microbiology is increasingly recognizing that the bacterial species normally and harmlessly found in humans also possess communication capabilities, an ability that may be key to some of our own normal physiological processes.[174]

» SOCIAL AMOEBAS

Another quite striking example of sociomicrobiology in action comes courtesy of our amoeboid friends: the case of a few species of social amoebas collectively called *slime molds* (please note that these are different from slime bacteria). These organisms are widely considered to be "on their way" to multicellularity, and in fact, they are truly multicellular, at least for part of their life cycle. To say that these are interesting organisms is an understatement. According to the scientist who knows them best, Princeton University's Dr. John Tyler Bonner,[175] slime molds were only recognized as a distinct species of organisms in the nineteenth century, most likely because in their unicellular form they were indistinguishable from any other amoeba species; conversely, in multicellular form they closely resembled individuals of a common species of bread molds. Of the several species of slime molds currently identified, the two best-studied genera are *Dictyostelium* and *Physarum*, each of which are represented by a number of distinct species.[176] Their colonies are formed by independent cells that are nonetheless able to cooperate, solve problems, and even seem to learn from experience.

Despite certain differences in the specifics, most slime molds follow a common "script." In times of plenty, the slime mold is not a slime mold at all: the individual amoeboid cells are free-living, eating bacteria at their leisure. It is only when food is scarce that the release of certain signaling molecules is triggered. These molecules somehow spread the word about the relative decrease in nutrient availability to the general amoeboid population. In turn, the collective response is an aggregation of many cells into an integrated mass, named a slug (come to think of it, this is not too different from quorum sensing).* Once aggregated, the cells differentiate into two main types: cells that form a stalk (about 20 percent of the total population) and cells that form a capsule that contains spores, which can be stored there for a period of time until the environmental conditions become favorable again. The whole structure is usually called the fruiting body. It is important to note that only the capsule cells participate in reproduction. Let me rephrase this: *a full fifth* of the total amoeboid population sacrifices

---

* There is an intriguing difference between *Dictyostelium* and *Physarum* in terms of their collective behavior. *Dictyostelium*'s cells retain their identity when together. Interestingly, *Physarum*'s individual amoebae fuse into a single mass, essentially forming a unicellular (or rather acellular) slug.

itself to help the rest of the fruiting body survive and reproduce. This fact illustrates an aspect of behavioral biology that puzzled scientists for a long time. This goes beyond simple cooperation, as it implies the "voluntary" sacrifice of a significant number of the population to help the rest of them survive and reproduce.

Slime molds are truly interesting organisms that can teach us a lot about "brainless" societies; these organisms have been found to possess quite unexpected behavioral capabilities![177] It is evident that we humans do not have a monopoly in the thinking department!*

SMALL BRAINS, BIG SOCIETIES

We have now seen examples of quite remarkable behaviors performed by true organismal societies, behaviors that seem to appear without the apparent presence or benefit of the "traditional hardware" in the form of anything that we'd recognize as a brain, or even a simple nervous system. These examples represent true steps towards multi-cellularity, the ultimate cell-to-cell cooperation scheme.

The appearance of nervous systems provided a way for multi-cellular organisms to communicate with each other to achieve tasks that no single individual would be able to achieve. Premier examples

---

* Again, please remember that these guys do not "think" in the traditional sense. There are many curious behavioral characteristics of these unusual organisms. I—very briefly—discussed the "intellectual life" of slime molds in more detail in Pagán, 2014, page 152: *"Another rather interesting trait of slime molds and related organisms is that they are capable of rather impressive feats traditionally thought to be limited to 'higher' animals. These include behaviors like problem-solving skills and the ability to learn. Amazingly, they also display the ability to anticipate environmental changes based on prior experience. Still, just like bacteria, the amoeba-like cells in slime molds do not possess an actual animal-like nervous system. If we think about it, from our admittedly biased perspective, the absence of a nervous system makes the behavioral repertoire of bacteria and slime molds even more astonishing. Curious creature as I am, this makes me ask myself, wouldn't it be interesting to find out whether bacterial populations have some form of self-awareness? Would it be possible at some fundamental level that slime molds, well, wonder?"*

of these include societies that are considered "superorganisms,"[178] because they express social behaviors that can be seen as analogous with the physiology of an individual organism.[179] The "poster kids" of superorganism societies include four main groups of animals: ants, termites, and some species of bees and wasps. In fact, they are not only the most social insects, they are the most social of all animals overall.[180] These organisms are quite remarkable in great part precisely because of their relatively small brains, which they more than compensate for with their sophisticated communication capabilities. Ants and bees have been used as examples of industriousness for centuries. However, naturalists only began to systematically study their communication capabilities in the late 1800s, mainly within the context of their foraging behavior.* For a long time, how social insects were able to communicate with each other remained a matter of speculation. It was only almost a century later that the visual behaviors of bees as well as chemical communication in ants were discovered and decoded, lines of research still under heavy scrutiny by scientists.

THE COLLECTIVE INTELLIGENCE OF ANTS

Just to give you an example of how intricate and frankly amazing insect societies are, let me start with the size of their brains. A typical ant has a brain containing close to 250,000 nerve cells—in comparison, a typical human brain has about 86,000,000 similar neurons. We will never know for sure what a particular ant is thinking, but it must not be much. However, thousands of ants working together can, depending on the species, find food, retain a memory of where the food is located, and return to the nest, leaving a chemical trail as directions for the other members of the colony.

---

* Yes, Charles Darwin was one of the naturalists that wondered about social insects.

Ants can also communicate with each other directly, again, via chemistry, using a variety of pheromones. They can also choose the optimal route by which to reach a location from among a couple of alternatives. Perhaps more surprisingly, ant colonies collectively engage in practices that one would usually associate with animals possessing considerably more brainpower. For example, they wage war with competing colonies, kidnap larvae of other ant species, and raise those young as slaves. They also practice agriculture, raising certain fungi species and harvesting them as a food source. And these are just a few examples of what these remarkable creatures can do![181]

Another slightly more technical term to refer to a superorganism is the label *eusocial*. The term itself means "true sociality," and to classify an organism species as such there are three nonnegotiable criteria: (1) the adult population displays division into two main castes, namely workers—that do not generally reproduce—and a reproductive caste (this is probably the defining characteristic of eusociality, as it is very unusual for an organism to forgo reproduction in favor of other members of its species); (2) at least two generations must coexist in the overall population; and (3) finally, the worker caste takes care of eggs and juvenile members of the colony.[182] A few biologists have proposed that humans themselves are eusocial, but the overwhelmingly general consensus is that this is a stretch—and I tend to agree with this assessment—we have not achieved true eusociality . . . yet. Anyway, in addition to ants, termites, wasps, and bees, there are at least four other types of organisms that rely on a eusocial lifestyle. One of these is a particular species of shrimp, *Synalpheus regalis*, described as eusocial in 1996. Since then, at least six other shrimp species have been determined to be eusocial. There is also at least one type of eusocial weevil (*Austroplatypus incompertus*; a curious scientific name, by the way!) as well as some species of aphids belonging to the *Pemphigus* genus. A particularly strange case of eusociality is represented by none other than a parasitic worm belonging to the genus *Himasthla*, described as recently as 2010.[183] This case is frankly surprising because these types of parasitic worms—essentially fluke flatworms—have been studied for the longest time, and we thought that we knew everything about them; clearly, we did not. This goes to further illustrate the fact that parasites are much more complex than most people give them credit for.

The general consensus seems to indicate that the "most" eusocial organisms are the almost fifty species of leafcutter ants. Alongside other eusocial insects like certain species of termites, these organisms literally invented agriculture about sixty million years ago. To this day, these societies actively culture various species of mold, which they consume for nourishment. Furthermore, certain other species invented the use of other animals for their own benefit; there are ant species that "raise" and protect insects like aphids. Ants do not eat them—rather, they "milk" the aphids and consume their sugary excretions. It is pretty evident that eusociality is a sound survival strategy. Even though roughly 2 percent of all insects are eusocial, they are the principal predators and scavengers of small prey, like other insects.[184] One may be excused for thinking that eusociality works very well in all these cases because, after all, invertebrate brains tend to be small. However, eusociality is not the exclusive purview of invertebrates. Prepare to be amazed!

» EUSOCIAL MAMMALS: MOLE RATS

Most interestingly, there are at least three species of truly eusocial* mammals, all of them examples of an unusual type: the mole rats. These were recognized as such relatively recently, in the 1970s. The

---

* One just has to be intrigued by how many different species of organisms, from bacteria to mammals, have developed behavioral tools that significantly enhance their survival chances. I have said so before, and I am not afraid of saying it again: biology is amazing!

story* of how these animals caught the attention of behavioral biologists is a beautiful example of the predictive power of evolutionary theories. In 1974, the entomologist Richard D. Alexander, currently an emeritus professor of biology at the University of Michigan, predicted that given a set of specific parameters, it was not out of the question that one or more mammalian species would evolve eusociality. Alexander developed a specific twelve-point model of the characteristics that such a mammal would have, all the way up to where they should live. He presented a series of lectures at various universities, including Northern Arizona University in Flagstaff. On that occasion, a member of the audience, the mammalogist Terry Vaughan, pointed out that the description of Alexander's hypothetical mammal perfectly fit the description of the naked mole rat, *Heterocephalus glaber*.[185] This sparked a renewed interest in this organism, particularly regarding their social behavior. Please recall that perhaps the most unusual behavioral trait of eusocial animals is the division of labor as far as reproduction is concerned. This means that almost invariably only one female reproduces, "assisted" (in the barest male sense) by a few males. The rest of the population does not reproduce. Let's perform a rather brief thought experiment: Can you imagine a human society where just a queen and a couple of consorts engage in reproductive activities? Me neither.

Probably the very first thing that one has to say about mole rats is that they are not attractive animals, not even a little bit. They lack the cuteness that we usually associate with small mammals. Naked mole rats are little, about four inches long at most, and besides their unusualness in terms of eusociality, they have caught the eye of scientists of many different disciplines for a variety of reasons, as they are unique to say the least. For example, they are cold-blooded, meaning that they cannot effectively control their body temperature in response to the ambient temperature. In this respect, they are more

---

* For a more extensive narrative of this story, please see Braude (1988).

like fish, amphibians, or reptiles than proper mammals. They also tend to be long-lived, especially for such a small organism; for example, a similarly sized mouse lives a couple of years at most, and this is in captivity, mind you; in the wild they probably live much shorter lives. In comparison, the longevity record for a naked mole rat seems to be about thirty years in the wild!

Part of the reason for their relatively long lives is that they are really tough critters. They are very resistant to painful stimuli, and due to a series of biochemical pathways very unusual for a mammal, they tolerate low oxygen levels remarkably well. In fact, a recent report[186] indicates that they can resist complete oxygen deprivation for up to eighteen minutes, recovering fully when normal air is available again. Once again, compared to mice, naked mole rats live sixteen minutes longer under such conditions. Scientists traced this ability to a certain biological trick that the mole rats use, namely, they can temporarily use an alternate biochemical cell pathway that does not require oxygen. Until very recently, it was even thought that these mammals do not develop cancer. It was subsequently found that they actually do, albeit at a much lower rate than what one would predict based on their life span. I am sure that these organisms have many tricks to teach us.

» THE BEE BALL

Honeybees are some of the most interesting animals on Earth. They are not only interesting—as master pollinators, they are essential for human society. Bees assist the reproductive activities of many plants that we consume. If bees were not around, our diet would be much less varied. In fact, several bee species are endangered. The reasons seem to be complex, and it is undeniably a topic worth studying. There are very few organisms without natural predators, and honeybees are frequently preyed upon by several wasp and hornet species. Bees are not pushovers though. They have the advantage of their social behavior, which, combined with their venomous sting, makes them formidable adversaries. However, they are not particularly aggressive, with the exception of certain breeds of Africanized honeybees, whose story is very well-known.* That being said, hornets tend to be bigger, meaner, and harder than any bee—in a literal sense. The hornets' exoskeleton generally makes them impervious to honeybee stings, and when they attack, they can devastate a beehive. Some accounts state that hornets can kill some forty bees a minute; therefore, a swarm of as few as ten hornets can do a lot of damage. Hornets feed on the bees' carcasses as well as on the bees' larvae. Truth be told, sometimes the bees are able to kill a few hornets by virtue of sheer numbers, but in this kind of fight, the odds are overwhelmingly in the hornets' favor. However, certain species of bees have evolved a kind of behavior that, frankly, boggles the mind. One cannot help but wonder how on Earth such a behavior developed. Alas, we do not know exactly how these kinds of bees came up with such an elaborate defensive mechanism, but at least we can see the results in action.

The Asian giant hornet (*Vespa mandarinia*) is a formidable creature. They grow up to two inches long, and have a mean streak. They are very aggressive, and in fact several people die each year due to *V. mandarinia* attacks.[187] This hornet is the main predator of *Apis cerana japonica*, the Japanese honeybee, a close cousin of the European honeybee, *Apis mellifera*. Japanese bees have evolved a strategy to deal with hornet attacks: they can form a "bee ball" around the

---

* Briefly, in Brazil during the 1950s, there were efforts to increase the honey productivity of honeybees. One of the approaches used was to produce hybrids of European and African honeybee varieties, because this latter variety displayed more honey production. The hybrid not only inherited the increased honey-producing capacity, but also inherited the aggressiveness and particularly short fuse of the African variety. One fine day, in a reality that rivaled the plot of a science fiction movie, several colonies broke through quarantine and escaped. The first of these hybrid colonies was detected in the United States for the first time in 1985, and they seem to be here to stay. They are in fact quite aggressive and have caused human fatalities. A really nice website with relevant information is found at: http://www.propacificbee.com/infographic/AHB/infographic.php.

attacker. This type of behavior was described as early as the 1920s,[188] but it was not until the 1980s that scientists were able to determine one of the reasons the Japanese bees do this.[189] When they let loose, a few bees restrain the hornet and keep it in place, while up to five hundred other bees surround the offending hornet. The bees then proceed to vibrate in a coordinated manner, which raises the temperature inside the bee ball to a few degrees above the tolerance limit of the hornet, but well within the bees' heat tolerance. So heat is one of the factors that kills the hornet, but more recently it has come to light that the levels of carbon dioxide and humidity also rise to lethal levels inside the ball, where the hornet is trapped; these multiple means conspire to kill the attacker.[190]

The Japanese honeybees do not only use this strategy against hornets, they also use it against wasps.[191] We are now beginning to understand the origins of this remarkable behavior, thanks in part to a rather interesting study that describes the differences between the nervous systems of bee species that engage in the formation of bee balls and bee species that do not.[192] I, for one, am extremely interested in knowing how such a sophisticated behavioral strategy evolved; it's bound to be a quite interesting story! Certainly the example of these bees is one of the most dramatic illustrations that cooperation—not size, not speed, not brute force—is perhaps the most efficient survival tactic of them all.

THE AGE OF WONDER[193]

As I said in the introduction, a normal-size book cannot cover every possible topic relevant to the overall theme of our exploration. There are so many things left to say, so many interesting organisms to talk about! There are many curiosities in nature left to discover, and this is especially true of the biological world. It is remarkable that we know so much more of what occurs at the core of the very stars we admire in the night sky than about the workings of a single cell. In my view, we are living through a modern Age of Wonder.* We still have the curiosity-driven mind of the eighteenth and nineteenth century naturalists, and yet we have the advantage of modern science and the technological advances that were not available to the early naturalists. We learn new and exciting things every single day, and I want to know so much more!

Humans are inquisitive and communicative organisms, and we are, of course, largely social. As we talked about earlier in this chapter, these traits are what have enabled us to survive for as long as we have. One of the most surefire ways to attract the attention of a fellow human being is to initiate a conversation, perhaps using the following words: "*Did you know . . . ?*" This is especially true of science enthusiasts like you and me. These three simple words will open a world of wonder for us. Aristotle probably said it best: "*Philosophy* [the love for knowledge] *starts in wonder and wonderment.*"

This innate curiosity that we humans possess is the basis for the semi-famous "gossip test," popularized by that scientists' scientist, the late Sir Francis Crick, of DNA fame. He wrote about this test in his book *What Mad Pursuit*,[194] a semi-biographical—and quite delightful—account of how most true scientists experience science.

---

* The Age of Wonder was a period during the eighteenth and nineteenth centuries when the practice of science in the modern sense blossomed and flourished. In those times, anyone who was so intellectually—and financially!—inclined could make scientific discoveries, and was able to acquire pretty much all the facts of nature as understood at the time. These men (and sadly, it was almost always men) called themselves *naturalists* regardless of whether their interests laid with the earth, the sky, or other areas of science entirely. It is no coincidence that many of our scientific giants lived their lives then. This era gave birth to a way of thinking about the natural world that scientists have adhered to ever since. This particular way of thinking accelerated the pace of scientific progress in an unprecedented way—and scientific progress, as we all can see, shows no sign of slowing down. Two excellent books that describe this era are: Snyder (2012) *The Philosophical Breakfast Club: Four Remarkable Friends Who Transformed Science and Changed the World*, and Holmes (2010) *The Age of Wonder: The Romantic Generation and the Discovery of the Beauty and Terror of Science*.

The gossip test as applied to the scientific endeavor is triggered when you just can't wait to share some curious fact, some strange organism, or, in the case of us research scientists, some exciting data. I absolutely relate to this feeling. It is the urge that makes you go into the hallway and grab (figuratively) the very first person that you run into and ask: *"Did you know . . . ?"* Or even sometimes *"Guess what?"* I speak from personal experience. In an academic environment, this behavior is common, kind of expected, and quite welcome. To some, it is endearing, as in a harmless example of eccentricity. However, it is much more than that, as this is an all-important urge. It is the proverbial spark that prompts the most important tool in the scientist's arsenal. The *"Did you know . . . ?"* question oftentimes leads to more questions. And it is only through more questions that scientific knowledge grows.

When I am thinking (and I do think a lot), I usually wonder about all the many things that can be initially prefaced by a *"Did you know . . . ?"* question. Nature, especially life, is deeply mysterious, even more so than we can possibly imagine. Just on this planet, we have described about two million species of living organisms, but all biologists agree that there are many more to discover. The current estimate for the total number of species on Earth ranges between eight and thirty million. It is no wonder that we discover new forms of life every single year. What else is there for us to look at? We are indeed living in exciting times.

In this book you have read about a small fraction of the wide variety of animals that use unusual strategies to survive. Some of the most surprising survival strategies that organisms display, if you ask me, are the ones that involve cooperation between two or more species, just like the ones you have read about in this very chapter. When thinking about all of these cooperative behaviors, there is one nagging thought, an idea in my mind that just does not let go. A thought that perfectly ends the journey that started at the beginning of this book. It is time to complete the circle. Here goes:

One of the most fundamental—perhaps the most fundamental— properties of biological life is cooperation.

In the opening chapters, we talked very briefly about what we know about the origin of life. And even though the discussion was relatively succinct, in reality, at the most elemental level, there is not much more to say about how life began. We are making progress in learning more about life's scientific mysteries, to be sure, but the fact remains that it boils down to chemistry. Oh, but what wonderful chemistry! We do know very precisely the type and relative amount of chemicals that make a particular human being, but try as we might, it is not a simple matter of mixing the right chemicals under the right conditions. You see, *life only occurs when inanimate molecules cooperate with each other.* We have yet to discover the fundamental physical laws that controlled the interaction of those early organic molecules that one fine day "decided" to work together toward a common goal by mastering the abilities of replication and metabolism. By necessity, there must have been various types of such groups of molecules around, because only then could selection processes enter into play, allowing the surviving ensembles of molecules to take over the planet in an invincible wave of what today we call life.

We can state the big question as: What fired us up?

This is a much deeper mystery than the very origin of species, what Darwin dubbed the "mystery of mysteries." Again, it is telling that we know more of what occurs at the very center of a star, nay, that we know more of what occurred a mere minute after the Big Bang* itself than we know about the fundamental properties of a living cell and what makes it tick. If life is ever going to understand itself (and, as far as we know, we are representing life as the only self-reflecting species on Earth), it is up to us as sentient beings to throw light onto what exactly life is, and how we can get to it from nonlife. Solving this puzzle will allow us to understand ourselves and the life around us a little more, perhaps—and hopefully—for the better.

---

* The very event that started the universe as we know it.

POSTSCRIPT

Did you know . . .

- that certain dolphin species may use puffer fish (a known source of the tetrodotoxin that we talked about in Chapter 1) to self-intoxicate? It seems that the dolphins get "high" by ingesting the substances secreted by the puffer fish. There is video evidence seeming to show that young dolphins even pass the fish around!*
- that there are quite a few species of social spiders?** And since we are at it, did you know that there is at least one species of vegetarian spider?[195]
- about a species of hornet that captures solar energy and converts it into usable bioelectricity?[196]
- about a certain ant species (*Formica yessensis*) that is able to build truly enormous communities? For example, in 1979 a supercolony of this ant species was found on the Japanese island of Hokkaido. It contained more than three hundred million

---

* Please see www.smithsonianmag.com/smart-news/dolphins-seem-to-use-toxic-pufferfish-to-get-high-180948219/. This has not been scientifically studied and at this point is anecdotal. But who knows? Incidentally, I am seriously thinking about exploring this example (and many other examples) of animal use of psychoactive substances as the topic of my next book (wink, wink).

** Please don't evolve wings. Pretty please, do not evolve wings!

individuals distributed throughout some forty-five thousand separate yet interconnected nests. Talk about a megapolis![197]

Are you curious yet about what other weird and wonderful things are out there? I hope so!

ACKNOWLEDGMENTS

Oh, my dear readers, if you only knew how this book came to be! That is a story for another time, though. For now I want to thank some very wonderful people who helped me make this project a reality.

First, I want to express my deepest appreciation to Mr. Glenn Yeffeth, Publisher and President of BenBella Books. Glenn took a chance on me and gave me wise—and, more importantly, right—advice. This book would not have been born without his help, and I will not forget the opportunity that he gave me.

I also want to thank the other members of the BenBella team with whom I worked directly: Scott Calamar, Jennifer Canzoneri, Sarah Avinger, Alicia Kania, Adrienne Lang, Monica Lowry, Rachel Phares, and Leah Wilson. It was a pleasure working with such a talented and professional team. As a big bonus, they are a lot of fun to work with too. Thank you one and all, guys!

There are two BenBella team members to whom I want to express my heartfelt appreciation for their invaluable help, guidance, and patience. They are two of my Jedi Masters (I will talk about a third one soon)—they were my editors for this book, and their names are Laurel Leigh and Alexa Stevenson. Alexa and Laurel, what can I tell you? Words are woefully inadequate to convey how grateful I am to you, so, simply stated: Thank you!

I want to thank my wife, Liza, for her help, suggestions, and above all, her love. She gave me her keen eye and her unrivaled common sense. Only Liza knows how to slow me down and keep me down to Earth in those times when I want to go "full scientific." This book is much better because of her. Also, I would be remiss if I didn't acknowledge our first 25 years together. I love you, my dream girl!

Vanessa, Reynaldo, Andy: What can I say? Because of you I try to be a better man and a better dad each and every day. I live for your love, for your hugs, for your smiles, and for your encouragement and understanding. It is an honor to be your dad, and I love you with all my heart.

Many good friends helped me through the various incarnations of this book. I want to highlight the help and friendship and express my special gratitude to my third Jedi Master, Ms. Sari J. Nichols. She helped me edit most of the book in its early incarnation and was my very own private writing teacher. My friends Dr. Cynthia Horton, Dr. Joy Fritschle, Mr. Martin Clemens, and Mr. John Jaksich all read and commented on at least one chapter. To all of you, thanks, and of course, any errors are mine and mine alone! My special thanks to my friend and fellow biologist Joshua Stuart Rose, who brought to my attention the case of certain eusocial parasitic worms, an interesting piece of information that made it in just in time before finishing the book.

A good scholar needs a good library to go to. I have said it before and I stand by this assertion. The library of my academic home, West Chester University, was invaluable when looking for information in many forms: books, original papers, electronic resources, and other miscellaneous material. My friend and colleague Dr. Walter Cressler is the library's "go-to" guy as far as biology is concerned. Thanks, Walt!

Finally, I want to express my special appreciation to Marie McNeely, PhD, and Peter Cawdron for your endorsements. I am humbled and honored by your kindness.

NOTES

[1] Please see Switek, *My Beloved Brontosaurus: On the Road with Old Bones, New Science, and Our Favorite Dinosaurs*, 2013. If you like dinosaurs, you'll love this book!

[2] Please see Xing and collaborators (2016).

[3] For a more extensive account of this historical period please refer to Emlen, *Animal Weapons: The Evolution of Battle*, 2015.

[4] There is a very lucid and layperson-friendly explanation of biological arms races in Dawkins, *River Out of Eden: A Darwinian View of Life*, 1996. If you are interested in a more technical account, please see Dawkins and Krebs (1979). I simply love the way Dawkins writes (about science, that is)!

[5] As we said before, an emerging threat for public health is the development of resistance to medications by infectious agents like viruses, bacteria, and parasites. For examples of this phenomenon related to parasites and viruses, please see Capewell and collaborators (2015) as well as Hoffman and collaborators (2015).

[6] Please see Pagán (2005).

[7] Please see Davis, *The Serpent and the Rainbow: A Harvard Scientist's Astonishing Journey into the Secret Societies of Haitian Voodoo, Zombies, and Magic*, 1985. For a more technical account please see Davis (1983).

[8] www.csicop.org/si/show/zombies_and_tetrodotoxin.

[9] For a more detailed narrative of this story see Brodie (2011).

[10] This idea is oftentimes expressed as "Bad genes or bad luck?" Please see Raup (1991, 1992).

[11] For more about the story of the snake and the newt, please see www.sciencedaily.com/releases/2008/03/080311075326.htm.

[12] Please see Zimmer (2011), "A Beautiful Web of Poison Extends A New Strand." blogs.discovermagazine.com/loom/2011/06/21/a-beautiful-web-of-poison-extends-a-new-strand.

[13] For a very readable and thorough exploration of the topic please see the book by Margulis and Sagan, *What Is Life?* (1995). By the way, this is one of the good books with *"What Is Life?"* as a title.

[14] For a short, nontechnical, and very good primer to viruses please see Crawford, *Viruses: A Very Short Introduction*, 2011.

[15] For example, an excellent basic reference about cells is: Allen and Cowling, *The Cell: A Very Short Introduction*, 2011.

[16] Again, please see baldscientist.wordpress.com/2016/02/22/the-improbability-of-you/.

[17] McAuliffe, *This Is Your Brain on Parasites: How Tiny Creatures Manipulate Our Behavior and Shape Society*, 2016.

[18] Incidentally, this account is adapted from my blog post: baldscientist. wordpress.com/2016/02/22/the-improbability-of-you/.

[19] Of course, in precise terms, only the first egg released has a 1/3,000 probability of being released. The second one has a probability of about 1/2,999, the third one 1/2,998, etc. However, this does not affect our argument too much.

[20] As I said, this is a *very* simplified definition. There are many ways in which a single gene can code for more than one specific protein. This fact merits a book of its own and in fact some of these have been written.

[21] Please see www.the-scientist.com/?articles.view/articleNo/40441/title/Human-Gene-Set-Shrinks-Again/.

[22] For a clear and detailed exposition of DNA, its history, and how it works, written by one of its central characters, please see Watson, *DNA: The Secret of Life*, 2009.

[23] If you want to know more about proteins, please see Tanford and Reynolds, *Nature's Robots: A History of Proteins*, 2004, and Cordes, *Hallelujah Moments: Tales of Drug Discovery*, 2014.

[24] Please see Dobzhansky (1973).

[25] Please see Lane and collaborators (2013).

[26] You can find the complete reasoning behind this figure in Garrett and Grisham (2016).

[27] Again, see Margulis and Sagan, *What Is Life?*

[28] I read this comparison for the first time in Toomey, *Weird Life: The Search for Life That Is Very, Very Different from Our Own*, 2013. You gotta read this book!

[29] Darwin wrote this phrase in a February 1871 letter to J. D. Hooker. Source: www.darwinproject.ac.uk.

[30] There is no shortage of good books about the search of the origin of life. Five of my favorites are: Cairns-Smith, *Seven Clues to the Origin of Life: A Scientific Detective Story*, 1990; Davies, *The Fifth Miracle: The Search for the Origin and Meaning of Life*, 1999; De Duve, *Vital Dust: The Origin and Evolution of Life on Earth*, 1995; Mesler and Cleaves, *A Brief History of Creation: Science and the Search for the Origin of Life*, 2015; and Schopf, *Life's Origin: The Beginnings of Biological Evolution*, 2002.

[31] This interesting story is narrated in detail in Chapter 10 of Lane, *Life Ascending: The Ten Great Inventions of Evolution*, 2010. You should read anything that Dr. Lane writes. He is that good.

[32] Even imagining this sound can be the source of a well-known fear in many people; this extreme fear has its own name: *astraphobia*.

[33] At any given time, there is an average of about ten lightning strikes per second, according to the World Wide Lightning Location Network (wwlln.net/new/map/). Using a little arithmetic, we'll find out that this is roughly equivalent to twenty-six million times per month!

[34] I saw this story in print for the first time on page 6 of a 2004 book by Walter Gratzer, Eurekas and Euphorias: *The Oxford Book of Scientific Anecdotes*, published by Oxford University Press. If you like scientific trivia, you'll love this book!

[35] On a more scientific note, the "taser" nickname as applied to electric eels recently appeared in a scholarly paper: Catania (2015c), "An Optimized Biological Taser: Electric Eels Remotely Induce or Arrest Movement in Nearby Prey."

[36] For a fascinating and very complete historical account of how Europeans learned about the electric catfish please see Piccolino and collaborators (2011).

[37] There are some internet sources that indicate that there is a species of electric catfish from China. However, I was unable to find a direct published reference about it.

[38] There are some general books out there that tell the tale of electric fish in a historical context. There are a couple of quite excellent books about it. One is Turkel, *Spark from the Deep: How Shocking Experiments with Strongly Electric Fish Powered Scientific Discovery*, 2013. This is a very reader-friendly book, quite a delightful read. If you want comprehensive, and I mean comprehensive, please see Finger and Piccolino, *The Shocking History of Electric Fishes: From Ancient Epochs to the Birth of Modern Neurophysiology*, 2011. This is a truly fascinating read and very well written too, but brace yourself: it is "information dense"!

[39] Please see Mandriota and collaborators (1965).

[40] Please see Mandriota and collaborators (1968).

[41] Please see Ebert and collaborators (2015) and Carvalho and White (2016).

[42] Please see Finger and Piccolino (2011), Chapter 2.

[43] Please see Finger and Piccolino (2011), Chapter 3. I'm telling you, this book is *comprehensive*!

[44] Please see Francis and Dingley (2015) and Tsoucalas and collaborators (2014).

[45] Please see Catania (2014, 2015a, b, c).

[46] Please see Wulf, *The Invention of Nature: The Adventures of Alexander Von Humboldt, the Lost Hero of Science*, 2016. Silly coincidence alert: the publisher of this book was John Murray Publishers, the same print that published Darwin's *Origin of Species*.

[47] Dr. Catania published a systematic study of such behaviors in 2016. Please see Catania, "Leaping Eels Electrify Threats" (2016), supporting Humboldt's account of a battle with horses. This article included a photograph of some of the experiments where he used a simulated predator head with LED indicators that lit up upon the eel's electric discharge. I found it somewhat silly that the predator that he chose to simulate was alligator- or crocodile-like. These reptiles would not attack eels from above the water as shown in the pictures. I know; I'm a nitpicker. Nonetheless, the experiments demonstrated this aspect of the eel's predatory behavior.

[48] Please see Nelson (2011).

[49] And who correctly speculated for the first time that these organs originated from muscle? The one and only Chuck Darwin!

[50] Which, by the way, is called the nicotinic acetylcholine receptor. Your muscles and mine contract because of the presence of this protein within our muscle cells. This receptor has a distinguished history in terms of neuropharmacology and biochemistry. In fact, it is found in nervous tissue as well, and many diseases and conditions like Alzheimer's disease, myasthenia gravis, and even some forms of epilepsy relate to this receptor. Also, these nicotinic receptors are thought to be related to brain functions like cognition, perception, and related phenomena. The story of how scientists studied the nature of the electric organ (both from rays and eels) and its relationship with the nicotinic receptors is fascinating. It includes characters from many biological and physical disciplines who used, of all things, snake venoms to determine how electric organs worked! A somewhat dated, yet still excellent, general article that describes this story in some detail is "Chemical Signaling in the Brain" (1993) by Jean-Pierre Changeux. If you want to go a little deeper in this topic please see Keesey (2005), Salazar and collaborators (2013), and Unwin (2013).

[51] For two very well told general renditions of this story please see Ashcroft (2012), and Turkel (2013), Chapter 3. For more detailed historical insights on Galvani, Volta, and the concept of animal electricity please see Bresadola (2008), Clower (1998), de Michelli (2011), and de Micheli-Serra (2012).

[52] Think Pikachu of Pokémon fame, for example, as well as some other comic superheroes. But in all seriousness, there have been a couple of bizarre reports of supposedly electrical insects. These reports are most likely exaggerations at best, but insects like these would be very interesting indeed. I saw this report on page 624 of the fascinating compendium *Incredible Life: A Handbook of Biological Mysteries* by William R. Corliss.

[53] For an especially readable article, please see Nelson (2011).

[54] Please see Hughes, *Sensory Exotica: A World Beyond Human Experience*, 2001, page 202.

[55] Please see Murray (1960).

[56] For example, see Tanaka and Reddien (2011), and Vervoort (2011).

[57] I describe Trembley's story with hydrae in some more technical detail: Pagán, *The First Brain: The Neuroscience of Planarians*, 2001.

[58] Some recent research has shed some light on hydra's extreme regeneration capacities. For example, please see www.rt.com/viral/376635-hydra-regeneration-memory-study/.

[59] I describe regeneration in several forms as well as planarians in a somewhat detailed way in Pagán, *The First Brain: The Neuroscience of Planarians*, 2014. I am sorry . . . I am really not trying to push my first book. I searched all over, and there is no popular science book dealing with regeneration. Perhaps I should write one (wink, wink).

[60] A millimeter is about the width of a fingernail, 1/1000 of a meter.

[61] Please see Powell (2010).

[62] This capacity was described in Piraino and collaborators (1996), and reviewed in Petralia and collaborators (2015).

[63] A thorough discussion of this exciting field is beyond the scope of this book, so please see Bargouth and collaborators (2015), Cohen and Venkatachalam (2014), Funk (2015), Levin (2012), and Sullivan and collaborators (2016).

[64] This research was described in Emmons-Bell and collaborators (2015).

[65] Please see www.scientificamerican.com/article/bioelectric-code/.

[66] An especially useful and very readable reference book is *Biology of Gila Monsters and Beaded Lizards* by Dr. Daniel D. Beck, published in 2005.

[67] Which includes three other additional species (some authors refer to them as subspecies): *H. exasperatum*, *H. alvarezi*, and *H. charlesbogerti*.

[68] If you want to go beyond Dr. Beck's book, a comprehensive bibliography organizing more than 1,400 *Heloderma* references by Beaman and collaborators (2006) can be found at www.cwu.edu/biology/sites/cts.cwu .edu.biology/files/helodermabibliography.pdf.

[69] Please see www.statemuseum.arizona.edu/exhibits/saguaro/ and Brown and Carmony, *Gila Monster: Facts and Folklore Of Americas Aztec Lizard*, 1999.

[70] Please see Hausman and Hillerman, *The Gift of the Gila Monster: Navajo Ceremonial Tales*, 1993, as cited in Beck (2005).

[71] But if you want to know more, please see Bhavsar and collaborators, "Evolution of Exenatide as a Diabetes Therapeutic" (2013), Cai and collaborators, "Long-Acting Preparations of Exenatide" (2013), and Furman, "The Development of Byetta (exenatide) from the Venom of the Gila Monster as an Anti-Diabetic Agent" (2012).

[72] Please see Fry and collaborators (2009). This paper describes really cool research that integrates anatomy, physiology, and biochemical pharmacology. Worth a read!

[73] Please see Fry and collaborators (2012). This is a thorough study including pretty much each and every venomous lizard family known so far.

[74] Please see www.scientificamerican.com/article/strange-but-true-drinking-too-much-water-can-kill/. For a more technical account please see Farrell and Bower (2003).

[75] Please see Fry (2016).

[76] From Dumbacher and Pruett-Jones (1996), as cited in Ligabue-Braun and Carlini (2015).

[77] Please see Nathanson and collaborators (1993), and Soloway (1976).

[78] Dr. Bryan G. Fry, as quoted in King (2013).

[79] Please see Calatayud and González (2003).

[80] Please see Bennett (1968).

[81] Please see Stix (2005).

[82] Please see Bhavsar and collaborators (2013)

[83] From Wilson, *The Future of Life*. Vintage Press, 2003: 121.

[84] From www.bbc.com/earth/story/20150422-the-worlds-most-poisonous-animal and animals.nationalgeographic.com/animals/amphibians/poison-frog/.

[85] Please see amphibiaweb.org/cgi/amphib_query?where-scientific_name=Pleurodeles+waltl.

[86] Please see Heiss and collaborators (2010).

[87] Please see Jared and collaborators (2015).

[88] Please see Chapter 6 in Margulis and Sagan, *Microcosmos: Four Billion Years of Microbial Evolution*, 1997. It is only fair to point out that the "Oxygen Holocaust" concept is under quite a bit of discussion and is by no means universally accepted among biologists. That being said, I think that the late Lynn Margulis was right more often than she was wrong, so I will side with her on this particular point.

[89] Please see Lane, *Oxygen: The Molecule That Made the World*, 2004. In fact, again, I recommend anything that Dr. Lane writes; I mean it. Trust me, you won't be disappointed.

[90] Please see Yager (1981). This is the original article describing remipedes for the first time.

[91] Incidentally, Dr. Yager discovered this very species in 1987.

[92] Please see von Reumont and collaborators (2014).

[93] Please see www.nature.com/news/first-venomous-crustacean-discovered-1.13985.

[94] I blogged an early form of this section; it is, of course, adapted and updated for this book. You can find the original post at baldscientist.wordpress.com/2014/08/08/a-curious-defense-mechanism-of-some-scorpions/. Also, please see mentalfloss.com/article/63467/toxic-tale-scorpion-can-make-two-kinds-venom.

[95] Please see Inceoglu and collaborators (2003).

[96] Please see Dutertre and collaborators (2014) and Prator and collaborators (2014).

[97] From Kutschera and Elliott (2010).

[98] Incidentally, the original meaning of the word "doctor" was "teacher." Please see baldscientist.wordpress.com/2013/10/22/what-does-the-title-doctor-really-mean/.

[99] Please see Yamasaki (2012).

[100] As well as in veterinary medicine; please see Sobczak and Kantyka (2014).

[101] Please see Hyson (2005).

[102] Check out www.populstat.info/Europe/francec.htm.

[103] Please see Weldon (2000).

[104] More about pitohuis in Dumbacher and collaborators (2004), and Dumbacher and collaborators (2008).

[105] For two accessible reviews on venomous mammals, please see Rode-Margono and Nekaris (2015) and Ligabue-Braun and collaborators (2012).

[106] Please find a more complete account at: www2.hawaii.edu/~bemorton/Neuroscience/Neurochemistry/Legend.html.

[107] Please see the note above.

[108] From Plotkin, *Tales of a Shaman's Apprentice: An Ethnobotanist Searches for New Medicines in the Amazon Rain Forest*, 1994: 126.

[109] Schutt, *Dark Banquet: Blood and the Curious Lives of Blood-Feeding Creatures*, 2008. This is an excellent, entertaining book on animals that feed on blood.

[110] In fact, one of the best articles about this animal is titled "Mad, Bad, and Dangerous to Know." Please see Nekaris and collaborators (2013).

[111] Please see Rode-Margono and Nekaris (2015); see previous mention.

[112] Please see Madani and Nekaris (2014).

[113] Please see Brodie (2009).

[114] For a short yet very accurate and accessible account of the ecological context of toxins and venoms, please see Brodie (2009).

[115] Please see Plotkin, *Medicine Quest: In Search of Nature's Healing Secrets*, 2000, Schmidt, *The Sting of the Wild*, 2016, and Wilcox, *Venomous: How Earth's Deadliest Creatures Mastered Biochemistry*, 2016. These are general books with a very good narrative. Worth a look!

[116] For a rather entertaining read on these matters please see Berman, *Zoom: How Everything Moves: From Atoms and Galaxies to Blizzards and Bees*, 2014.

[117] Check out discovermagazine.com/2009/dec/16-the-brain-what-is-speed-of-thought.

[118] But just in case you are interested, here are some very good general articles on jellyfish stings and venoms: Cegolon and collaborators (2013), Fenner (1998), and Montgomery and collaborators (2016). An especially thorough and interesting resource on jellyfish venoms is Jouiaei and collaborators: "Ancient Venom Systems: A Review on Cnidaria Toxins" (2015).

[119] This fascinating research was published in a very readable article: Nüchter and collaborators, "Nanosecond-Scale Kinetics of Nematocyst Discharge" (2006).

[120] Certain fungi are able to shoot their spores at bullet-like speeds. There is a class of trap-jaw ants that close their jaws so fast that they can use this to jump over certain distances. Then there are the ultrafast tongues of frogs, toads, and chameleons. If you want to see many of these examples in more detail, please see Nüchter and collaborators (2016) and Sakes and collaborators (2016).

[121] For more information on how nudibranchs escape, steal, and reuse cnidarian stings please see Greenwood (2009).

[122] This discussion is adapted from a highly readable and relevant article, Patek: "The Most Powerful Movements in Biology" (2015).

[123] Again, please see Patek (2015).

[124] In fact, one of my favorite papers on mantis shrimp (in great part because of its title) is "The Stomatopod Dactyl Club: A Formidable Damage-Tolerant Biological Hammer" by Weaver and collaborators, published in 2012. This paper examines the hammer of a particular mantis shrimp species, *Odontodactylus scyllarus*, and its fascinating material and biomechanical properties.

[125] For a representative example please see: phys.org/news/2016-05-mantis-shrimp-ultra-strong-materials.html.

126 I wrote an early version of some of these ideas at baldscientist.wordpress
.com/2012/09/15/she-says-periwinkle-i-say-blue-the-invasion-of-the-
tetrachromats/.

127 Please see Thoen and collaborators: "A Different Form of Color Vision in
Mantis Shrimp" (2014).

128 For accessible perspectives on these extraordinary visual features of the
mantis shrimp, please see Hurtley: "Living Technicolor" (2014); Land and
Osorio: "Extraordinary Color Vision" (2014); and Pennisi: "Extraordinary
Eyes" (2012).

129 Please see Hann, Pagán, Eterović: "The Alpha-Conotoxins GI and MI
Distinguish Between the Nicotinic Acetylcholine Receptor Agonist Sites
While SI Does Not" (1994).

130 For excellent general reviews on these behaviors please see Olivera and
collaborators (2015), Olivera and Cruz (2001), and Salisbury (2010).

131 As told by Olivera and Cruz, "Conotoxins, in Retrospect." *Toxicon*. 39(1)
(2001): 7–14.

132 Other preeminent conotoxins researchers to date hail from Australia,
particularly Drs. Paul Alewood, David Craik, and Richard Lewis. I am
very grateful to Dr. Richard Lewis for kindly sending me several important
references.

133 Please see Armishaw and collaborators (2005), Bingham and collaborators
(2014), Halai and Craik (2009), and Robinson and Norton (2014).

134 Please see Safavi-Hemami and collaborators (2015). Specialized insulin is
used for chemical warfare by fish-hunting cone snails.

135 Please see Safavi-Hemami and collaborators (2015).

136 Please see Terlau and Olivera (2004).

137 Please see www.imdb.com/title/tt0598023/. I learned of this example from
the 2012 *Scientific American* web article by Daisy Yuhas: "Healing the Brain
with Snail Venom," found at www.scientificamerican.com/article/healing-
the-brain-with-snail-venom/.

138 Please see http://cinemorgue.wikia.com/wiki/Jordan_Murphy.

139 Please see http://jurassicpark.wikia.com/wiki/Lindstradt_Air_Gun.

140 Reported in Sarramegna (1965). Downloaded from www.spc
.int/DigitalLibrary/Doc/FAME/Reports/Sarramegna_65_
PoisonousGastropods_Conidae_NeWCal.pdf. This is a particularly good
reference that reviews several key properties of cone snails and their history.

[141] Please see Nagami (2004), page 60.

[142] To give you a better idea of his distinguished career, he's been publishing papers on the field since the 1950s and his latest paper was just published in 2014. Also in 2014, he published an authoritative book just about cone snails: *"Conus" of the Southeastern United States and Caribbean*, published by Princeton University Press, already recognized as a classic on the field.

[143] He was quite the interesting character too. Dr. Endean was one of the first scientists who called themselves toxinologists, with toxins and venoms as their main research topic. In his career, he studied the venom of fish, jellyfish, cone snails of course, and many others. For a more complete account of Dr. Endean's career please see Hawgood (2006).

[144] For more information on these interesting spiders please see Ariki and collaborators (2016), Zobel-Thropp and collaborators (2014), Gilbert and Rayor (1985), and Suter and Stratton (2009).

[145] For more technical accounts of the Burgess Shale, its contributions to paleontology, and its relationship with the Cambrian Explosion please see Budd (2008, 2013) and Conway Morris (2003).

[146] For a more detailed account of the discovery of the Burgess Shale please see two excellent general books: *Wonderful Life: The Burgess Shale and the Nature of History*, 1990, by the late Stephen Jay Gould (how I do miss that guy's writing!) and *The Crucible of Creation*, 1998, by Simon Conway Morris, the very paleontologist who described the weird prehistoric organism that we'll talk about in this section. Conway Morris also published a slightly more technical account of Walcott's discoveries. Please see Conway Morris (2009).

[147] Please see Zhuravlev and Riding, *The Ecology of the Cambrian Radiation*, 2000.

[148] Please see Smith and Ortega-Hernández: "Hallucigenia's Onychophoran-like Claws and the Case for Tactopoda" (2014).

[149] Please see Jönsson and collaborators: "Tardigrades Survive Exposure to Space in Low Earth Orbit" (2008), and Jönsson: "Tardigrades as a Potential Model Organism in Space Research" (2007).

[150] Please see Fortey, *Horseshoe Crabs and Velvet Worms: The Story of the Animals and Plants That Time Has Left Behind*, 2012.

[151] For a photograph showing some colorful velvet worms please see Blaxter and Sunnucks: Velvet worms (2011).

[152] There is a picture of a stunning one at: www.doc.govt.nz/Documents/conservation/native-animals/invertebrates/peripatus-facts-brochure.pdf.

[153] Please see Mayer and collaborators (2009, 2010).

[154] Please see Reinhard and Rowell, "Social Behaviour in an Australian Velvet Worm, *Euperipatoides rowelli*" (2005).

[155] To learn more about how these guys eat, please see Mayer and collaborators, "Capture of Prey, Feeding, and Functional Anatomy of the Jaws in Velvet Worms" (2015).

[156] For a thorough analysis of this topic, please see Weiss (2010).

[157] For more detailed and technical explanations of these principles please see Biernaskie and West (2015) and West and collaborators (2007).

[158] To learn more about symbiosis, please see Archibald, *One Plus One Equals One: Symbiosis and the Evolution of Complex Life*, 2014.

[159] There are two excellent general books if you want to learn more about Dr. Margulis's life in science: Sagan, (editor), *Lynn Margulis: The Life and Legacy of a Scientific Rebel*, 2012, and Margulis, Sagan, *Dazzle Gradually: Reflections on the Nature of Nature*, 2007. If you want a little more information on the scientific specifics please see Margulis and Bermudes: "Symbiosis as a Mechanism of Evolution: Status of Cell Symbiosis Theory" (1985); Margulis and collaborators: "Community Living Long Before Man: Fossil and Living Microbial Mats and Early Life" (1986); and Guerrero and collaborators: "Symbiogenesis: The Holobiont as a Unit of Evolution" (2013). Sadly, near the end of her productive scientific life, she became associated with controversial topics that affected her credibility.

[160] For an extended account of this theory in nontechnical language please see Margulis and Sagan, *Microcosmos: Four Billion Years of Microbial Evolution*, 1997; Margulis, *Symbiotic Planet: A New Look at Evolution*, 1999; and Margulis and Sagan, *Acquiring Genomes: A Theory of the Origin of Species*, 2003. For more technical information, please see Archibald (2015), Martin and collaborators (2015), and Zimorski and collaborators (2014).

[161] For a couple of reader-friendly accounts of this episode of the history of science, please see Jeon (1995a) and Sagan and Margulis (1987).

[162] Please see Leedale (1967).

[163] Some general papers that tell this wonderful story in detail are Jeon and Jeon (1976) and Jeon (1987, 1995a, b).

[164] For more information on other examples of endosymbiosis please see Blackstone (2016), Gentil and collaborators (2017), Geurts and collaborators (2016), and López-García and Moreira (2015).

[165] For a general account of this story, please see Dworkin: "Lingering Puzzles About Myxobacteria: How These Social Bacteria Form Multicellular Structures, Move and Change Shape, and Affect Soil Ecology Are Key Unsolved Questions" (2007).

[166] Described in detail in Dworkin: *The Myxobacteria: New Directions in Studies of Procaryotic Development* (1972).

[167] Please see Parsek and Greenberg, "Sociomicrobiology: the connections between quorum sensing and biofilms" (2005).

[168] In brief, Wilson's sociobiology proposed that social behavior results from evolutionary forces, a notion that is pretty much accepted today, but when it was proposed in the 1970s caused quite a bit of controversy inside and outside academia, in no small part for political instead of scientific reasons. Please see Wilson, *Sociobiology: The New Synthesis*, 2000. This could be the topic of a whole new book, and of course, we cannot cover this issue here. If you are interested, please see www.nybooks.com/articles/1975/11/13/against-sociobiology/.

[169] For an excellent general article on the topic please see Zimmer: "A Weakness in Bacteria's Fortress" (2015). If you want a little more depth, please see Greenberg, "The New Science of Sociomicrobiology and the Realm of Synthetic and Systems Ecology" (2011).

[170] Please see Xavier, "Sociomicrobiology and Pathogenic Bacteria" (2016).

[171] Based on Diggle and collaborators (2007).

[172] Prokaryote intercellular cooperation is not limited to bacteria. The "other" prokaryotes, the archaea, also display this type of behavior, although it is lesser understood. For examples, please see Fröls, "Archaeal Biofilms: Widespread and Complex" (2013), and Orell and collaborators, "Archaeal Biofilms: The Great Unexplored" (2013).

[173] Please see Finch and collaborators, "Quorum Sensing: A Novel Target for Anti-infective Therapy" (1998). But, as you may imagine, the idea of targeting quorum sensing as a way to attack bacteria came from nature first! Please see Bauer and Robinson, "Disruption of Bacterial Quorum Sensing by Other Organisms" (2002).

[174] For an excellent general book on the topic, please see Dietert, *The Human Superorganism: How the Microbiome Is Revolutionizing the Pursuit of a Healthy Life*, 2016.

[175] Please see Bonner (1949, 2008).

[176] For a couple of short, very reader-friendly articles on *Dictyostelium* and *Physarum*, please see Alim and collaborators (2013), and Fets and collaborators (2010).

[177] For an excellent general book on slime molds, please see Bonner (2008). For a more detailed description of slime mold biology, please see Nakagaki (2001), Nakagaki and collaborators (2000), and Saxe (1999).

[178] The concept of colonies as organisms seems to have originated with a paper published in 2011: Wheeler, W. M., "The Ant-Colony as an Organism." *Journal of Morphology* 22, no. 2 (1911): 307–325.

[179] Please see Wilson, *The Insect Societies*, 1971.

[180] Please see Wilson and Gómez Durán, *Kingdom of Ants: José Celestino Mutis and the Dawn of Natural History in the New World*, 2010.

[181] For more information, please see Chittka and Leadbeater (2005), Jackson and Ratnieks (2006), and Leadbeater and Chittka (2007).

[182] Please see Hölldobler and Wilson, *The Superorganism: The Beauty, Elegance, and Strangeness of Insect Societies*, 2008.

[183] Please see Newey and Keller, "Social Evolution: War of the Worms" (2010).

[184] For two excellent general references on eusociality please see Hölldobler and Wilson, *The Superorganism: The Beauty, Elegance, and Strangeness of Insect Societies*, 2008, and Hölldobler and Wilson, *The Leafcutter Ants: Civilization by Instinct*, 2010. For a couple of more specialized works, please see Wilson and Hölldobler (2005) and Queller and Strassmann (2003).

[185] Please see Bromham and Harvey (1996), Griffin (2008), and Jarvis (1981).

[186] Please see Storz and McClelland (2017).

[187] Please see Sugahara and collaborators (2012).

[188] Please see Tokuda, *Studies on the Honey Bee, with Special Reference to the Japanese Honey Bee*, 1924.

[189] Please see Ono and collaborators, "Heat Production by Balling in the Japanese Honeybee, *Apis cerana japonica* as a Defensive Behavior Against the Hornet, *Vespa simillima xanthoptera*" (1987).

[190] Please see Sugahara and collaborators, "Differences in Heat Sensitivity Between Japanese Honeybees and Hornets Under High Carbon Dioxide and Humidity Conditions Inside Bee Balls" (2012), and Sugahara and Sakamoto, "Heat and Carbon Dioxide Generated by Honeybees Jointly Act to Kill Hornets" (2009).

191 Please see Ken and collaborators (2005) and Tan and collaborators (2010).

192 Please see Ugajin and collaborators (2012).

193 This section contains elements of a post from my blog: baldscientist .wordpress.com/2015/08/14/did-you-know/.

194 Please see Crick, *What Mad Pursuit: A Personal View of Scientific Discovery*, 1990. This is a wonderful little book!

195 Yep, there is at least one of those! Please see Jackson (2007).

196 Please see Ishai (2004) and Plotkin and collaborators (2010).

197 Please see Hölldobler and Wilson, *The Leafcutter Ants: Civilization by Instinct*, 2010.

BIBLIOGRAPHY AND FURTHER READING

INTRODUCTION

Emlen, D. J. *Animal Weapons: The Evolution of Battle*, Reprint Edition. New York: Picador, 2015.

Jahren, H. *Lab Girl*. New York: Alfred A. Knopf, 2016.

Riskin, D. *Mother Nature Wants to Kill You*. New York: Simon & Schuster, 2015.

CHAPTER 1: THE *E* WORD

Did you know . . . that there is a new type of bear that is a hybrid of polar bears and grizzlies? Polar bears and grizzlies have long been known to mate and breed in captivity, but no one knew of such pairings in the wild until about ten years ago, when hunters began to find bears with characteristics of both. These hybrids are named depending on the father: If the dad is a grizzly, the cub is a "grolar bear." If the dad is a polar bear, the cub is called a "pizzly bear." Although there is some controversy about the specific reason that these bears have begun appearing in the wild, most scientists theorize that climate change causes more brown bears to migrate north and more polar bears to migrate south, increasing the chances of papa bear and mama bear finding each other. Interestingly, some hybrids have been found to have 75 percent of the genes of one bear type versus the other. This can only mean one thing: the hybrids are breeding! This is really exciting, because we are observing evolution in real time.[*]

Brodie III, E. D. "Patterns, Process, and the Parable of the Coffeepot Incident: Arms Races Between Newts and Snakes from Landscapes to Molecules," in *In the Light of Evolution: Essays from the Laboratory and Field*, ed. Jonathan Losos. Englewood, CO: Roberts & Company, 2011.

Capewell, P. et al. "A Co-evolutionary Arms Race: Trypanosomes Shaping the Human Genome, Humans Shaping the Trypanosome Genome." *Parasitology* 142, Suppl 1 (2015): S108–19.

Davis, W. *The Serpent and the Rainbow: A Harvard Scientist's Astonishing Journey into the Secret Societies of Haitian Voodoo, Zombis, and Magic*. New York: Simon and Schuster, 1985.

Davis, E. W. "The Ethnobiology of the Haitian Zombi." *Journal of Ethnopharmacology* 9, no. 1 (1983): 85–104.

Dawkins, R. *River Out of Eden: A Darwinian View of Life*. Science Masters Series. New York: Basic Books, 1995.

Dawkins, R. and Krebs, J. R. "Arm Races Between and Within Species." *Proceedings of the Royal Society of London* B 205 (1979): 489–511.

Emlen, D. J. *Animal Weapons: The Evolution of Battle*, Reprint Edition. New York: Picador, 2015.

Hoffmann, H. H. et al. "Interferons and Viruses: An Evolutionary Arms Race of Molecular Interactions." *Trends Immunol* 36, no. 3 (2015): 124–38.

Pallen, M. *The Rough Guide to Evolution*. London: Rough Guides, 2009.

Pagán, O. R. *Synthetic Local Anesthetics as Alleviators of Cocaine Inhibition of the Human Dopamine Transporter*. PhD dissertation. Ithaca, NY: Cornell University, 2005.

Raup, D. "Extinction: Bad Genes or Bad Luck?" *New Scientist* 131, no. 1786 (1991): 46–9.

---

[*] Please see https://sites.google.com/site/sciencefairhybridanimals/hybrids/bears/grolar-bear and http://www.slate.com/articles/news_and_politics/explainer/2010/05/pizzly_bears.html.

Raup, D. *Extinction: Bad Genes or Bad Luck?* New York: W. W. Norton & Company, 1992.

Switek, B. *My Beloved Brontosaurus: On the Road with Old Bones, New Science, and Our Favorite Dinosaurs.* New York: Scientific American / Farrar, Straus and Giroux, 2013.

Xing, L. et al. "A Feathered Dinosaur Tail with Primitive Plumage Trapped in Mid-Cretaceous Amber." *Current Biology* 26, no. 24 (2016): 3352–60.

CHAPTER 2: THE LANGUAGE OF LIFE

Did you know . . . that plants eat light and air? Granted, we all know that green plants need light to thrive, using an exquisite process called photosynthesis. Plants capture energy from light and transform that energy into the chemical energy that every living being needs (you will recall the discussion of adenosine triphosphate, or ATP). What many of us may not realize is that the matter that plants need to grow does not come from the soil. In fact, let's suppose that you plant a seed into a pot with a known amount of soil (say, five pounds) and lovingly tend to the growing plant until it is a tall shrub. If you transplant the shrub to the ground and weigh the soil in the pot, you will find still very close to five pounds of dirt. Where did the matter of the plant come from? Well, it came from the carbon dioxide in the air, the very same carbon dioxide that many organisms including ourselves breathe out as part of our physiology. How many different things the language of life can write!*

Allen, T. and Cowling G. *The Cell: A Very Short Introduction.* New York: Oxford University Press, 2011.

Cairns-Smith, A. G. *Seven Clues to the Origin of Life: A Scientific Detective Story (Canto).* New York: Cambridge University Press, 1990.

Cordes, E. H. *Hallelujah Moments: Tales of Drug Discovery.* New York: Oxford University Press, 2014.

---

* The details of this can be found in pretty much any good biology textbook.

Crawford, D. H. *Viruses: A Very Short Introduction*. New York: Oxford University Press, 2011.

Davies, P. *The Fifth Miracle: The Search for the Origin and Meaning of Life*. New York: Simon & Schuster, 1999.

De Duve, C. *Vital Dust: The Origin and Evolution of Life on Earth*, Revised Edition. New York: Basic Books, 1995.

Dobzhansky, T. "Nothing in Biology Makes Sense Except in the Light of Evolution." *The American Biology Teacher* 35, no. 3 (1973): 125–9.

Garrett, R. H. and Grisham C. M. *Biochemistry*. Boston, MA: Cengage Learning, 2016.

Lahav, N. *Biogenesis: Theories of Life's Origin*. New York: Oxford University Press, 1999.

Lane, N. *Life Ascending: The Ten Great Inventions of Evolution*. New York: W. W. Norton & Company, 2010.

Lane, N. et al. "Energy, Genes and Evolution: Introduction to an Evolutionary Synthesis." *Philosophical Transactions of the Royal Society of London* B Biol Sci. 368, no. 1622:20120253 (2013).

Margulis, L. and Sagan, D. *What is Life?* Oakland, CA: University of California Press, 1995.

McAuliffe, K. *This Is Your Brain on Parasites: How Tiny Creatures Manipulate Our Behavior and Shape Society*. Boston, MA: Eamon Dolan / Houghton Mifflin Harcourt, 2016.

Mesler, B. and Cleaves II, H. J. *A Brief History of Creation: Science and the Search for the Origin of Life*. New York: W. W. Norton & Company, 2015.

Schopf, J. W., ed. *Life's Origin: The Beginnings of Biological Evolution*. Oakland, CA: University of California Press, 2002.

Schrödinger, E. *What Is Life?* New York: Cambridge University Press, 2012.

Tanford, C. and Reynolds, J. *Nature's Robots: A History of Proteins*. New York: Oxford University Press, 2004.

Toomey, D. *Weird Life: The Search for Life That Is Very, Very Different from Our Own*. New York: W. W. Norton & Company, 2013.

Watson, J. D. *DNA: The Secret of Life*, Reprint Edition. New York: Knopf, 2009.

CHAPTER 3: IT ALL STARTS WITH A SPARK

Did you know . . . that there are animals that "grow up" but never "mature?"* Metamorphosis is a very familiar phenomenon. For example, caterpillars turn into pupae and then into butterflies; in fact, most insect species develop in such a way. Most amphibians like frogs and salamanders hatch from eggs as aquatic tadpoles, which subsequently change their shape (metamorphose) into adult form. However, there is a type of amphibian, the axolotl (*Ambystoma mexicanum*), that does not follow this sequence. These animals display *neoteny*, meaning that they retain juvenile characteristics while reaching sexual maturity; in essence, "metamorphosis, interrupted." Metamorphosis is a rather mysterious phenomenon. This warrants yet another "Did you know . . . ?" There are some organisms that undergo metamorphosis and then are able to remember what they learned while in the larval stage. These include a few types of insects and even some newts! What is remarkable about this is that during metamorphosis the nervous system is entirely reorganized. So how are these memories preserved? Metamorphosis is heavily influenced by bioelectricity. I fully expect quite a few research breakthroughs in this area in the future.**

Ashcroft, F. *The Spark of Life: Electricity in the Human Body.* New York: W. W. Norton & Company, 2012.

Barghouth, P. G. et al. "Bioelectrical Regulation of Cell Cycle and the Planarian Model System." *Biochimica et Biophysica Acta* 1848, no. 10 Pt B (2015): 2629–37.

Bresadola, M. "Animal Electricity at the End of the Eighteenth Century: The Many Facets of a Great Scientific Controversy." *Journal of the History of the Neurosciences* 17, no. 1 (2008): 8–32.

---

* I am not talking about boyfriends or husbands here . . .
** Please see Pai and collaborators (2015); Levin (2014); De Loof and collaborators (2014); Tully and collaborators (1994); Blackiston and collaborators (2008); and Hershkowitz and Samuel (1973).

Carvalho, M. R. and White, W. T. "*Narcine Baliensis*, a New Species of Electric Ray from Southeast Asia (Chondrichthyes: Torpediniformes)." *Zootaxa* 4127, no. 1 (2016):149–60.

Catania, K. C. "Leaping Eels Electrify Threats, Supporting Humboldt's Account of a Battle with Horses." *Proceedings of the National Academy of Sciences* USA 113, no. 25 (2016): 6979–84.

Catania, K. C. "Electric Eels Concentrate Their Electric Field to Induce Involuntary Fatigue in Struggling Prey." *Current Biology* 25, no. 22 (2015): 2889–98.

Catania, K. C. "Electric Eels Use High-Voltage to Track Fast-Moving Prey." *Nature Communications* 6, no. 8638 (2015b).

Catania, K. C. "An Optimized Biological Taser: Electric Eels Remotely Induce or Arrest Movement in Nearby Prey." *Brain, Behavior and Evolution* 86, no. 1 (2015): 38–47.

Changeux, J. P. "Chemical Signaling in the Brain." *Scientific American* 269, no. 5 (1993): 58–62.

Clower, W. T. "The Transition from Animal Spirits to Animal Electricity: A Neuroscience Paradigm Shift." *Journal of the History of the Neurosciences* 7, no. 3 (1998): 201–18.

Cobo, B. *Historia del Nuevo Mundo*. Manuscrito en Lima, Perú (1653).

Cohen, A. E. and Venkatachalam, V. "Bringing Bioelectricity to Light." *Annual Review of Biophysics* 43 (2014): 211–32.

Corliss, W. R. *Incredible Life: A Handbook of Biological Mysteries*. The Sourcebook Project, Ann Arbor, MI: 1981.

De Micheli, A. "On the First Studies of Electrophysiology." *Archivos de Cardiología de México* 81, no. 4 (2011): 337–42.

De Micheli-Serra, A. et al. "How Electricity Was Discovered and How It Is Related to Cardiology." *Archivos de Cardiología de México* 82, no. 3 (2012): 252–9.

Ebert, D. A. et al. "*Tetronarce cowleyi*, Sp. Nov., a New Species of Electric Ray from Southern Africa (Chondrichthyes: Torpediniformes: Torpedinidae)." *Zootaxa* 3936, no. 2 (2015): 237–50.

Emmons-Bell, M. et al. "Gap Junctional Blockade Stochastically Induces Different Species-Specific Head Anatomies in Genetically Wild-Type *Girardia dorotocephala* Flatworms." *International Journal of Molecular Sciences* 16, no. 11 (2015): 27865–96.

Finger, S. and Piccolino, M. *The Shocking History of Electric Fishes: From Ancient Epochs to the Birth of Modern Neurophysiology.* New York: Oxford University Press, 2011.

Francis, J. and Dingley, J. "Electroanaesthesia—From Torpedo Fish to TENS." *Anaesthesia* 70, no. 1 (2015): 93–103.

Funk, R. H. "Endogenous Electric Fields as Guiding Cue for Cell Migration." *Frontiers in Physiology* 6, no. 143 (2015).

Gilbert, W. *On the Magnet and Magnetic Bodies, and on That Great Magnet the Earth.* London: Peter Short, 1600.

Gratzer, W. *Eurekas and Euphorias: The Oxford Book of Scientific Anecdotes.* New York: Oxford University Press, 2004.

Hughes, H. C. *Sensory Exotica: A World Beyond Human Experience.* Cambridge, MA: Bradford Books, 2001.

Keesey, J. "How Electric Fish Became Sources of Acetylcholine Receptor." *Journal of the History of the Neurosciences* 14, no. 2 (2005): 149–64.

Levin, M. "Molecular Bioelectricity in Developmental Biology: New Tools and Recent Discoveries: Control of Cell Behavior and Pattern Formation by Transmembrane Potential Gradients." *BioEssays* 34, no. 3 (2012): 205–17.

Mandriota, F. J. et al. "Classical Conditioning of Electric Organ Discharge Rate in Mormyrids." *Science* 150 (1965): 1740–42.

Mandriota, F. J. et al. "Avoidance Conditioning of the Rate of Electric Organ Discharge in Mormyrid Fish." *Animal Behaviour* 16, no. 4 (1968): 448–55.

Murray, R. W. "Electrical Sensitivity of the Ampullae of Lorenzini." *Nature* 187: 957 (1960).

Nelson, M. E. "Electric Fish." *Current Biology* 21, no. 14 (2011): R528–9.

Pagán, O. R. *The First Brain: The Neuroscience of Planarians.* New York: Oxford University Press, 2014.

Petralia, R. S. et al. "Aging and Longevity in the Simplest Animals and the Quest for Immortality." *Ageing Research Reviews* 16 (2014): 66–82.

Piccolino, M. et al. "Discovering the African Freshwater 'Torpedo': Legendary Ethiopia, Religious Controversies, and a Catfish Capable of Reanimating Dead Fish. *Journal of the History of the Neurosciences* 20, no. 3 (2011): 210–35.

Piraino, S. et al. "Reversing the Life Cycle: Medusae Transforming into Polyps and Cell Transdifferentiation in *Turritopsis Nutricula* (Cnidaria, Hydrozoa)." *Biological Bulletin* 190 (1996): 302–12.

Powell, K. "Masters of Regeneration." *HHMI Bulletin* 23, no. 3 (2010): 28–48.

Salazar, V. L. et al. "The Energetics of Electric Organ Discharge Generation in Gymnotiform Weakly Electric Fish." *Journal of Experimental Biology* 216, Pt. 13 (2013): 2459–68.

Sullivan, K. G. et al. "Physiological Inputs Regulate Species-Specific Anatomy During Embryogenesis and Regeneration." *Communicative & Integrative Biology* 9, no 4 (2016): e1192733.

Tanaka, E. M. and Reddien, P. W. "The Cellular Basis for Animal Regeneration." *Developmental Cell* 21, no. 1 (2011): 172–85.

Tsoucalas, G. et al. "The 'Torpedo' Effect in Medicine." *International Maritime Health* 65, no. 2 (2014): 65–7.

Turkel, W. J. *Spark from the Deep: How Shocking Experiments with Strongly Electric Fish Powered Scientific Discovery*. Baltimore, MD: Johns Hopkins University Press, 2013.

Unwin, N. "Nicotinic Acetylcholine Receptor and the Structural Basis of Neuromuscular Transmission: Insights from Torpedo Postsynaptic Membranes." *Quarterly Review of Biophysics* 46, no. 4 (2013): 283–322.

Vervoort, M. "Regeneration and Development in Animals." *Biological Theory* 6 (2011): 25–35.

Wulf, A. *The Invention of Nature: The Adventures of Alexander Von Humboldt, the Lost Hero of Science*. London: John Murray Publishers, 2016.

CHAPTER 4: UNUSUAL SUSPECTS

Did you know... that one of the most effective medications against high blood pressure was inspired by a snake's venom? The Brazilian pit viper (*Bothrops jararaca*) accounts for close to 90 percent of snake bites in South America. One of the main symptoms of its bite is an extreme and therefore very dangerous decrease in blood pressure. Understandably, the venom of this snake was screened to find substances capable of lowering blood pressure. Inspired by some compounds from the pit viper's venom, the first of such drugs (Captopril) was brought to market in the 1980s. Over time, the discovery and development of Captopril paved the way for additional compounds that have proven useful as antihypertensive agents, all of this due to the keen observations and carefully designed experiments that integrated the specific expertise, ideas, and abilities of a wide variety of researchers. This story is an example of biomedical science at its best. Moreover, it is not the only example of medicines from venoms. We have barely scratched the surface of this rich source of potential medicinal drugs!*

Beaman, K. R. et al. "The Beaded Lizard (*Heloderma Horridum*) and Gila Monster (*Heloderma Suspectum*): A Bibliography of the Family Helodermatidae." *Smithsonian Herpetological Information Service*, no. 136 (2006).

Beck, D. D. *Biology of Gila Monsters and Beaded Lizards*. Oakland, CA: University of California Press, 2005.

Bennett, A. E. "The History of the Introduction of Curare into Medicine." *Anesthesia & Analgesia* 47, no. 5 (1968): 484–92.

Bhavsar, S. et al. "Evolution of Exenatide as a Diabetes Therapeutic." *Current Diabetes Review* 9, no. 2 (2013): 161–93.

Bhavsar, S., Mudaliar, S., and Cherrington, A. "Evolution of Exenatide as a Diabetes Therapeutic." *Current Diabetes Reviews* 9, no. 2 (2013): 161–93.

Brodie III, E. D. "Toxins and Venoms." *Current Biology* 19, no. 20 (2009): R931–5.

Brown, D. E. and Carmony, N. E. *Gila Monster: Facts & Folklore of America's Aztec Lizard*. Salt Lake City: University of Utah Press, 1999.

Cai, Y. et al. "Long-Acting Preparations of Exenatide." *Drug Design, Development and Therapy* 7 (2013): 963–70.

---

* Please see Smith and Vane (2003); King (2013); Downey (2008); and Waheed and collaborators (2017).

Calatayud, J. and González, A. "History of the Development and Evolution of Local Anesthesia Since the Coca Leaf." *Anesthesiology* 98, no. 6 (2003): 1503–8.

Dumbacher, J. P. et al. "Phylogeny of the Avian Genus Pitohui and the Evolution of Toxicity in Birds." *Molecular Phylogenetics and Evolution* 49, no. 3 (2008): 774–81.

Dumbacher, J. P. et al. "Melyrid Beetles (Choresine): A Putative Source for the Batrachotoxin Alkaloids Found in Poison-Dart Frogs and Toxic Passerine Birds." *Proceedings of the National Academy of Sciences* USA 101, no. 45 (2004): 15857–60.

Dumbacher, J. P. and Pruett-Jones, S. "Avian Chemical Defense." *Current Ornithology* 13, Nolan Jr., V. and Ketterson, E. D., eds., New York: Plenum Press, 1996: 137–174.

Dutertre, S. et al. "Evolution of Separate Predation- and Defence-Evoked Venoms in Carnivorous Cone Snails." *Nature Communications* 5, no. 3521 (2014).

Farrell, D. J. and Bower, L. "Fatal Water Intoxication." *Journal of Clinical Pathology* 56, no. 10 (2003): 803–4.

Fry, B. G. *Venom Doc: The Edgiest, Darkest, Strangest Natural History Memoir Ever Made*. New York: Arcade Publishing, 2016.

Fry, B. G. et al. "A Central Role for Venom in Predation by *Varanus Komodoensis* (Komodo Dragon) and the Extinct Giant *Varanus (Megalania) Priscus*." *Proceedings of the National Academy of Sciences* USA 106, no. 22 (2009): 8969–74.

Fry, B. G. et al. "The Structural and Functional Diversification of the Toxicofera Reptile Venom System. *Toxicon* 60, no. 4 (2012): 434–48.

Furman, B. L. "The Development of Byetta (Exenatide) from the Venom of the Gila Monster as an Anti-Diabetic Agent." *Toxicon* 59, no. 4 (2012): 464–71.

Heiss, E. et al. "Hurt Yourself to Hurt Your Enemy: New Insights on the Function of the Bizarre Antipredator Mechanism in the Salamandrid *Pleurodeles Waltl*." *Journal of Zoology* 280, no. 2 (2010): 156–162.

Hyson, J. "Leech Therapy: A History." *Journal of the History of Dentistry* 53, no. 1 (2005): 25–7.

Inceoglu, B. et al. "One Scorpion, Two Venoms: Prevenom of *Parabuthus Transvaalicus* Acts as an Alternative Type of Venom with Distinct Mechanism of Action." *Proceedings of the National Academy of Sciences* USA 100, no. 3 (2003): 922–7.

Jared, C. et al. "Venomous Frogs Use Heads as Weapons." *Current Biology* 25, no. 16 (2015): 2166–70.

King, G. F. "Venoms as a Platform for Human Drugs: Translating Toxins into Therapeutics." *Expert Opinion on Biological Therapy* 11, no. 11 (2011): 1469–84.

Kutschera, U. and Elliott, J. M. "Charles Darwin's Observations on the Behaviour of Earthworms and the Evolutionary History of a Giant Endemic Species from Germany, *Lumbricus Badensis* (Oligochaeta: Lumbricidae)." *Applied and Environmental Soil Science* 2 (2010): 1–11.

Lane, N. *Oxygen: The Molecule that Made the World*. New York: Oxford University Press, 2004.

Ligabue-Braun, R. and Carlini, C. R. "Poisonous Birds: A Timely Review." *Toxicon* 99 (2015): 102–8.

Madani, G. and Nekaris, K. A. "Anaphylactic Shock Following the Bite of a Wild Kayan Slow Loris (*Nycticebus Kayan*): Implications for Slow Loris Conservation." *Journal of Venomous Animals and Toxins Including Tropical Diseases* 20, no. 1 (2014): 43.

Nathanson, J. A. et al. "Cocaine as a Naturally Occurring Insecticide." *Proceedings of the National Academy of Sciences* USA 90, no. 20 (1993): 9645–8.

Nekaris, K. A. et al. "Mad, Bad and Dangerous to Know: The Biochemistry, Ecology and Evolution of Slow Loris Venom." *Journal of Venomous Animals and Toxins Including Tropical Diseases* 19, no. 1 (2013): 21.

Plotkin, M. J. *Tales of a Shaman's Apprentice: An Ethnobotanist Searches for New Medicines in the Amazon Rain Forest*. New York: Penguin Books, 1994.

Plotkin, M. J. *Medicine Quest: In Search of Nature's Healing Secrets*. New York: Penguin Books, 2000.

Prator, C. A. et al. "Venom Variation During Prey Capture by the Cone Snail, *Conus Textile*." *PLoS One* 9, no. 6 (2014): e98991.

Rode-Margono, J. E. and Nekaris, K. A. "Cabinet of Curiosities: Venom Systems and Their Ecological Function in Mammals, with a Focus on Primates." *Toxins* (Basel) 7, no. 7 (2015): 2639–58.

Schmidt, J. O. *The Sting of the Wild*. Baltimore, MD: Johns Hopkins University Press, 2016.

Schutt, B. *Dark Banquet: Blood and the Curious Lives of Blood-Feeding Creatures*. New York: Harmony Books, 2008.

Sobczak, N. and Kantyka, M. "Hirudotherapy in Veterinary Medicine." *Annals of Parasitology* 60, no. 2 (2014): 89–92.

Soloway, S. B. "Naturally Occurring Insecticides." *Environmental Health Perspectives* 14 (1976): 109–17.

Stix, G. "A Toxin Against Pain." *Scientific American* 292, no.4 (2005): 70–5.

Von Reumont, B. M. et al. "The First Venomous Crustacean Revealed by Transcriptomics and Functional Morphology: Remipede Venom Glands Express a Unique Toxin Cocktail Dominated by Enzymes and a Neurotoxin." *Molecular Biology and Evolution* 31, no. 1 (2014): 48–58.

Weldon, P. J. "Avian Chemical Defense: Toxic Birds Not of a Feather." *Proceedings of the National Academy of Sciences* USA 97, no. 24 (2000): 12948–9.

Wilcox, C. *Venomous: How Earth's Deadliest Creatures Mastered Biochemistry.* New York: Scientific American/Farrar, Straus and Giroux, 2016.

Wilson, E. O. *The Future of Life.* New York: Vintage Press, 2003.

Yager, J. "Remipedia, a New Class of Crustacea from a Marine Cave in the Bahamas." *Journal of Crustacean Biology* 1, no. 3 (1981): 328–33.

Yager, J. "*Speleonectes Tulumensis*, n. sp. (Crustacea: Remipedia) from Two Anchialine Cenotes of the Yucatan Peninsula, Mexico." *Stygologia* 3, no. 2 (1987): 160–6.

Yamasaki, A. "Leeches in Medicine: Friend or Foe?" *TuftScope* 11, no. 2 (2012): 1–4.

CHAPTER 5: THE FAST AND THE HANGRY

Did you know . . . that there is such a thing as a tentacled snake? You read it right; the tentacled snake (*Erpeton tentaculatum*) is a proper snake, meaning that it has no appendages at all, except for two finger-like structures that sprout from its face. These snakes are exclusively aquatic and fish feeders and were first described in the 1800s. The specific nature of their "tentacles" was a source of puzzlement and fascination in equal measure. Some ideas on what their tentacles were used for included camouflage, as worm-like lures to trap fish, as mere ornaments, and as sensory organs. This last one turns out to be correct. These organs are extremely sensitive receptors of water pressure. This means that they are able to detect very slight disturbances in water flow, which may signal the presence of a fish, a very useful ability if you live in murky waters that prevent you from seeing much. Tentacled snakes are described as "sit-and-wait" predators. They do not actively seek prey, but rather they wait for the prey to come to them. They combine their ability to detect fish with their strike speed; they strike at an average speed of 25 milliseconds (0.025 seconds). Interestingly, baby snakes do not have to be taught either to sit and wait or to strike, as it seems that these are innate abilities. There you have it: an ultrasensitive motion sensor coupled with an ultrarapid strike. This snake definitively has what it takes to survive![*]

Ariki, N. K. et al. "Characterization of Three Venom Peptides from the Spitting Spider *Scytodes Thoracica.*" *PLoS One* 11, no. 5 (2016): e0156291.

Armishaw, C. J. and Alewood, P. F. "Conotoxins as Research Tools and Drug Leads." *Current Protein & Peptide Science* 6, no. 3 (2005): 221–40.

Berman, R. *Zoom: How Everything Moves: From Atoms and Galaxies to Blizzards and Bees.* New York: Little, Brown and Company, 2014.

Bingham, J-P et al. "Conotoxins," in *Manual of Security Sensitive Microbes and Toxins,* ed. D. Liu. Boca Raton, FL: CRC Press, 2014.

Blaxter, M. and Sunnucks, P. "Velvet Worms." *Current Biology* 21, no. 7 (2011): R238–40.

Budd, G. E. "At the Origin of Animals: The Revolutionary Cambrian Fossil Record." *Current Genomics* 14, no. 6 (2013): 344–54.

Budd, G. E. "The Earliest Fossil Record of the Animals and Its Significance." *Philosophical Transactions of the Royal Society of London* B Biol Sci. 363, no. 1496 (2008): 1425–34.

Catania, K. C. "The Brain and Behavior of the Tentacled Snake." *Annals of the New York Academy of Sciences* 1225 (2011): 83–9.

[*] Please see Catania (2010, 2011, 2012a, 2012b) and Catania and collaborators (2010).

Catania, K. C. "Born Knowing: Tentacled Snakes Innately Predict Future Prey Behavior." *PLoS One* 5, no. 6 (2010): e10953.

Catania, K. C. et al. "Function of the Appendages in Tentacled Snakes (*Erpeton Tentaculatus*)." *Journal of Experimental Biology* 213, no. 3 (2010): 359–67.

Catania, K. C. "Evolution of Brains and Behavior for Optimal Foraging: A Tale of Two Predators." *Proceedings of the National Academy of Sciences* USA 109 Suppl 1 (2012a): 10701–8.

Catania, K. C. "Tactile Sensing in Specialized Predators—From Behavior to the Brain." *Current Opinion in Neurobiology* 22, no. 2 (2012b): 251–8.

Cegolon, L. et al. "Jellyfish Stings and Their Management: A Review." *Marine Drugs* 11, no. 2 (2013): 523–50.

Conway-Morris, S. "The Cambrian 'Explosion' of Metazoans and Molecular Biology: Would Darwin Be Satisfied?" *International Journal of Developmental Biology* 47, no. 7–8 (2003): 505–15.

Conway-Morris, S. "Walcott, the Burgess Shale and Rumours of a Post-Darwinian World." *Current Biology* 19, no. 20 (2009): R927–31.

Conway-Morris, S. *The Crucible of Creation: The Burgess Shale and the Rise of Animals*. New York: Oxford University Press, 1998.

Fenner, P. J. "Dangers In the Ocean: The Traveler and Marine Envenomation. I. Jellyfish." *Journal of Travel Medicine* 5, no. 3 (1998): 135–41.

Fortey, R. *Horseshoe Crabs and Velvet Worms: The Story of the Animals and Plants that Time Has Left Behind*. New York: Knopf, 2012.

Gilbert, C. and Rayor, L. S. "Predatory Behavior of Spitting Spiders (Araneae: Scytodidae) and the Evolution of Prey Wrapping." *Journal of Arachnology* 13 (1985): 231–41.

Greenwood, P. G. "Acquisition and Use of Nematocysts by Cnidarian Predators." *Toxicon* 54, no. 8 (2009): 1065–70.

Gould, S. J. *Wonderful Life: The Burgess Shale and the Nature of History*. New York: W. W. Norton & Company, 1989.

Halai, R. and Craik, D. J. "Conotoxins: Natural Product Drug Leads." *Natural Product Reports* 26, no. 4 (2009): 526–36.

Hann, R. M., Pagán, O. R., and Eterović, V. A. "The Alpha-Conotoxins GI and MI Distinguish Between the Nicotinic Acetylcholine Receptor Agonist Sites While SI Does Not." *Biochemistry* 33, no. 47 (1994): 14058–63.

Hawgood, B. J. "The Marine Biologist—Bob Endean." *Toxicon* 48, no. 7 (2006): 768–79.

Hurtley, S. "Living Technicolor." *Science* 343, no. 6169 (2014): 350.

Jönsson, K. I. et al. "Tardigrades Survive Exposure to Space in Low Earth Orbit." *Current Biology* 18, no. 17 (2008): R729–31.

Jönsson, K. I. "Tardigrades as a Potential Model Organism in Space Research." *Astrobiology* 7, no. 5 (2007): 757–66.

Jouiaei, M. et al. "Ancient Venom Systems: A Review on Cnidaria Toxins." *Toxins* (Basel) 7, no. 6 (2015): 2251–71.

Kohn, A. J. *"Conus" of the Southeastern United States and Caribbean.* Princeton, NJ: Princeton University Press, 2014.

Land, M. F. and Osorio, D. "Extraordinary Color Vision." *Science* 343, no. 6169 (2014): 381–2.

Mayer, G. et al. "Capture of Prey, Feeding, and Functional Anatomy of the Jaws in Velvet Worms (Onychophora)." *Integrative & Comparative Biology* 55, no. 2 (2015): 217–27.

Mayer, G. et al. "A Revision of Brain Composition in Onychophora (Velvet Worms) Suggests that the Tritocerebrum Evolved in Arthropods." *BMC Evolutionary Biology* 10 (2010): 255.

Mayer, G. and Whitington, P. M. "Neural Development in Onychophora (Velvet Worms) Suggests a Step-Wise Evolution of Segmentation in the Nervous System of Panarthropoda." *Developmental Biology* 335, no. 1 (2009): 263–75.

Montgomery, L. et al. "To Pee, or Not to Pee: A Review on Envenomation and Treatment in European Jellyfish Species." *Marine Drugs* 14, no. 7 (2016): 127.

Nagami, P. *Bitten: True Medical Stories of Bites and Stings.* New York: St. Martin's Press, 2004.

Nüchter, T. et al. "Nanosecond-Scale Kinetics of Nematocyst Discharge." *Current Biology* 16, no. 9 (2006): R316–8.

Olivera, B. M. et al. "Prey-Capture Strategies of Fish-Hunting Cone Snails: Behavior, Neurobiology and Evolution." *Brain, Behavior and Evolution* 86, no. 1 (2015): 58–74.

Olivera, B. M. and Cruz, L. J. "Conotoxins, in Retrospect." *Toxicon* 39, no. 1 (2001): 7–14.

Patek, S. N. "Materials Science. Biomimetics and Evolution." *Science* 345, no. 6203 (2014): 1448–9.

Patek, S. N. "The Most Powerful Movements in Biology." *American Scientist* 103 (2015): 330–7.

Patek, S. N. et al. "From Bouncy Legs to Poisoned Arrows: Elastic Movements in Invertebrates." *Journal of Experimental Biology* 214, pt. 12 (2011): 1973–80.

Patek, S. N. et al. "Comparative Spring Mechanics in Mantis Shrimp." *Journal of Experimental Biology* 216, pt. 7 (2013): 1317–29.

Patek, S. N. and Caldwell, R. L. "Extreme Impact and Cavitation Forces of a Biological Hammer: Strike Forces of the Peacock Mantis Shrimp *Odontodactylus Scyllarus*." *Journal of Experimental Biology* 208, pt. 19 (2008): 3655–64.

Pennisi, E. "Extraordinary Eyes." *Science* 335, no. 6073 (2012): 1163.

Reinhard, J. and Rowell, D. M. "Social Behaviour in an Australian Velvet Worm, *Euperipatoides Rowelli* (Onychophora: Peripatopsidae)." *Journal of Zoology* 267, no. 1 (2005): 1–7.

Robinson, S. D. and Norton, R. S. "Conotoxin Gene Superfamilies." *Marine Drugs* 12, no. 12 (2014): 6058–101.

Safavi-Hemami, H. et al. "Specialized Insulin Is Used for Chemical Warfare by Fish-Hunting Cone Snails." *Proceedings of the National Academy of Sciences* USA 112, no. 6 (2015): 1743–8.

Sakes, A. et al. "Shooting Mechanisms in Nature: A Systematic Review." *PLoS One* 11, no. 7 (2016): e0158277.

Sarramegna, R. "Poisonous Gastropods of the Conidae Family Found in New Caledonia and the Indo-Pacific." *South Pacific Commission Technical Paper* No. 144 (1965).

Smith, M. R. and Ortega-Hernández, J. "Hallucigenia's Onychophoran-like Claws and the Case for Tactopoda." *Nature* 514, no. 7522 (2014): 363–6.

Suter, R. B. and Stratton, G. E. "Spitting Performance Parameters and Their Biomechanical Implications in the Spitting Spider, *Scytodes Thoracica*." *Journal of Insect Science* 9, no. 62 (2009): 1–15.

Terlau, H. and Olivera, B. M. "Conus Venoms: A Rich Source of Novel Ion Channel-Targeted Peptides." *Physiological Reviews* 84, no. 1 (2004): 41–68.

Thoen, H. H. et al. "A Different Form of Color Vision in Mantis Shrimp." *Science* 343, no. 6169 (2014): 411–3.

Weaver, J. C. et al. "The Stomatopod Dactyl Club: A Formidable Damage-Tolerant Biological Hammer." *Science* 336, no. 6086 (2012): 1275–80.

Zobel-Thropp, P. A. et al. "Spit and Venom from Scytodes Spiders: A Diverse and Distinct Cocktail." *Journal of Proteome Research* 13, no. 2 (2014): 817–35.

Zhuravlev, A. and Riding, R. *The Ecology of the Cambrian Radiation*. New York: Columbia University Press, 2000.

CHAPTER 6: THE VERY BEST SURVIVAL TACTIC OF THEM ALL

Did you know . . . that there are various types of ants whose societies include specialized workers who sacrifice themselves in the most extreme way, literally by exploding? In the 1970s, scientists discovered the first example of such an ant species (*Camponotus saundersi*). Their workers possess enlarged digestive glands, and when a predator attacks, the ant contracts its gland in such a way that it bursts, showering the attacker with a corrosive and sticky solution. There are various additional species of ants that use essentially the same strategy with slight variations. Talk about cooperation!*

Alim, K. et al. "Physarum." *Current Biology* 23, no. 24 (2013): R1082–3.

Archibald, J. M. "Endosymbiosis and Eukaryotic Cell Evolution." *Current Biology* 25, no. 19 (2015): R911–21.

Archibald, J. *One Plus One Equals One: Symbiosis and the Evolution of Complex Life*. Oxford, NY: Oxford University Press, 2014.

Bauer, W. D. and Robinson, J. B. "Disruption of Bacterial Quorum Sensing by Other Organisms." *Current Opinion in Biotechnology* 13, no. 3 (2002): 234–7.

---

* Please see Davidson and collaborators (2011) and Jones and collaborators (2004).

Biernaskie, J. M. and West, S. A. "Cooperation, Clumping and the Evolution of Multicellularity." *Proceedings of the Royal Society of London* B 282, no. 1813 (2015): 1075.

Blackstone, N. W. "An Evolutionary Framework for Understanding the Origin of Eukaryotes." *Biology* (Basel) 5, no. 2 (2016): E18.

Bonner, J. T. "The Social Amoebae." *Scientific American* 180, no. 6 (1949): 44–7.

Bonner J, T. *The Evolution of Culture in Animals*. Princeton, NJ: Princeton University Press, 1983.

Bonner, J. T. *The Social Amoebae: The Biology of Cellular Slime Molds*. Princeton, NJ: Princeton University Press, 2008.

Braude, S. "The Predictive Power of Evolutionary Biology and the Discovery of Eusociality in the Naked Mole-Rat." *NCSE Reports* 17, no. 4 (1998): 12–15.

Bromham, L. and Harvey, P. H. "Naked Mole-Rats on the Move: Behavioural Ecology." *Current Biology* 6, no. 9 (1996): 1082–3.

Chittka, L. and Leadbeater, E. "Social Learning: Public Information in Insects." *Current Biology* 15, no. 21 (2005): R869–71.

Crick, F. *What Mad Pursuit: A Personal View of Scientific Discovery*. New York: Basic Books, 1990.

Dietert, R. *The Human Superorganism: How the Microbiome Is Revolutionizing the Pursuit of a Healthy Life*. New York: Dutton, 2016.

Diggle, S. P. et al. "Quorum Sensing." *Current Biology* 17, no. 21 (2007): R907–10.

Dworkin, M. "Lingering Puzzles About Myxobacteria: How These Social Bacteria Form Multicellular Structures, Move and Change Shape, and Affect Soil Ecology Are Key Unsolved Questions." *Microbe* 2, no. 1 (2007): 18–24.

Dworkin, M. "The Myxobacteria: New Directions in Studies of Procaryotic Development." *CRC Critical Reviews in Microbiology* 2 (1972): 435–52.

Fets, L. et al. "Dictyostelium." *Current Biology* 20, no. 23 (2010): R1008–10.

Finch, R. G. et al. "Quorum Sensing: A Novel Target for Anti-infective Therapy." *Journal of Antimicrobial Chemotherapy* 42, no. 5 (1998): 569–71.

Fröls, S. "Archaeal Biofilms: Widespread and Complex." *Biochemical Society Transactions* 41, no. 1 (2013): 393–8.

Gentil, J. et al. "Review: Origin of Complex Algae by Secondary Endosymbiosis: A Journey Through Time." *Protoplasma* (e-pub ahead of print; 2017).

Geurts, R. et al. "What Does It Take to Evolve a Nitrogen-Fixing Endosymbiosis?" *Trends in Plant Science* 21, no. 3 (2016): 199–208.

Gould, S. J. "Evolution in Fact and Theory." In *Hen's Teeth and Horse's Toes*, pp. 253–62. New York: W. W. Norton & Company, 1983.

Greenberg, E. P. "The New Science of Sociomicrobiology and the Realm of Synthetic and Systems Ecology." In *The Science and Applications of Synthetic and Systems Biology: Workshop Summary* from the Institute of Medicine (US) Forum on Microbial Threats. Washington, DC: National Academies Press (US), page 213 (2011).

Griffin, A. S. "Naked Mole-Rat." *Current Biology* 18, no. 18 (2008): R844–5.

Guerrero, R. et al. "Symbiogenesis: The Holobiont as a Unit of Evolution." *International Microbiology* 16, no. 3 (2013): 133–43.

Hölldobler, B. and Wilson, E. O. *The Superorganism: The Beauty, Elegance, and Strangeness of Insect Societies*. New York: W. W. Norton & Company, 2008.

Hölldobler, B. and Wilson, E. O. *The Leafcutter Ants: Civilization by Instinct*. New York: W. W. Norton & Company, 2010.

Holmes, R. *The Age of Wonder: The Romantic Generation and the Discovery of the Beauty and Terror of Science*. New York: Vintage, 2010.

Ito, F. et al. "What Is for Dinner? First Report of Human Blood in the Diet of the Hairy-Legged Vampire Bat *Diphylla Ecaudata*." *Acta Chiropterologica* 18, no. 2 (2016): 509–15.

Jackson, D. E. and Ratnieks, F. L. "Communication in Ants." *Current Biology* 16, no. 15 (2006): R570–4.

Jarvis, J. U. "Eusociality in a Mammal: Cooperative Breeding in Naked Mole-Rat Colonies." *Science* 212, no. 4494 (1981): 571–3.

Jeon, K. W. and Jeon, M. S. "Endosymbiosis in Amoebae: Recently Established Endosymbionts Have Become Required Cytoplasmic Components." *Journal of Cellular Physiology* 89, no. 2 (1976): 337–44.

Jeon, K. W. "Change of Cellular 'Pathogens' into Required Cell Components." *Annals of the New York Academy of Sciences* 503 (1987): 359–71.

Jeon, K. W. "Bacterial Endosymbiosis in Amoebae." *Trends in Cell Biology* 5, no. 3 (1995a): 137–40.

Jeon, K. W. "The Large, Free-Living Amoebae: Wonderful Cells for Biological Studies." *The Journal of Eukaryotic Microbiology* 42, no. 1 (1995b): 1–7.

Ken, T. et al. "Heat-Balling Wasps by Honeybees." *Naturwissenschaften* 92, no. 10 (2005): 492–5.

Leadbeater, E. and Chittka, L. "Social Learning in Insects—From Miniature Brains to Consensus Building." *Current Biology* 17, no. 16 (2007): R703–13.

Leedale, G. F. "Euglenida-Euglenophyta." *Annual Review of Microbiology* 21 (1967): 31–48.

Ligabue-Braun, R. et al. "Venomous Mammals: A Review." *Toxicon* 59, no. 7–8 (2012): 680–95.

López-García, P. and Moreira, D. "Open Questions on the Origin of Eukaryotes." *Trends in Ecology and Evolution* 30, no. 11 (2015): 697–708.

Margulis, L. *Symbiotic Planet: A New Look at Evolution*. New York: Basic Books, 1999.

Margulis, L. et al. "Community Living Long Before Man: Fossil and Living Microbial Mats and Early Life." *Science of the Total Environment* 56 (1986): 379–97.

Margulis, L. and Sagan, D. *Microcosmos: Four Billion Years of Microbial Evolution*. Oakland, CA: University of California Press, 1997.

Margulis, L. and Sagan, D. *Acquiring Genomes: A Theory of the Origin of Species*. New York: Basic Books, 2003.

Margulis, L. and Sagan, D. *Dazzle Gradually: Reflections on the Nature of Nature*. White River Junction, VT: Chelsea Green Publishing, 2007.

Margulis, L. and Bermudes, D. "Symbiosis as a Mechanism of Evolution: Status of Cell Symbiosis Theory." *Symbiosis* 1 (1985): 101–24.

Martin, W. F. et al. "Endosymbiotic Theories for Eukaryote Origin." *Philosophical Transactions of the Royal Society of London* B Biol Sci. 370, no. 1678:20140330 (2015).

Miller, S. "Production of Amino Acids Under Possible Primitive Earth Conditions." *Science* 117, no. 3046 (1953): 528–9.

Nakagaki, T. "Smart Behavior of True Slime Mold in a Labyrinth." *Research in Microbiology* 152, no. 9 (2001): 767–70.

Nakagaki, T. et al. "Maze-Solving by an Amoeboid Organism." *Nature* 407, no. 470 (2000).

Newey, P. and Keller, L. "Social Evolution: War of the Worms." *Current Biology* 20, no. 22 (2010): R985–7.

Ono, M. et al. "Heat Production by Balling in the Japanese Honeybee, *Apis Cerana Japonica* as a Defensive Behavior Against the Hornet, *Vespa Simillima Xanthoptera* (Hymenoptera: Vespidae)." *Experientia* 43 (1987): 1031–2.

Orell, A. et al. "Archaeal Biofilms: The Great Unexplored." *Annual Review of Microbiology* 67 (2013): 337–54.

Pagán, O. R. *The First Brain: The Neuroscience of Planarians*. New York: Oxford University Press, 2014.

Parsek, M. R. and Greenberg, E. P. "Sociomicrobiology: The Connections Between Quorum Sensing and Biofilms." *Trends in Microbiology* 13, no. 1 (2005): 27–33.

Queller, D. C. and Strassmann, J. E. "Eusociality." *Current Biology* 13, no. 22 (2003): R861–3.

Sagan, D., ed. *Lynn Margulis: The Life and Legacy of a Scientific Rebel*. White River Junction, VT: Chelsea Green Publishing, 2012.

Sagan, D. and Margulis, L. "Bacterial Bedfellows: A Microscopic Ménage à Trois May Be Responsible for a Major Step in Evolution." *Natural History* 96, no. 3 (1987): 26–33.

Saxe, C. L. "Learning From the Slime Mold: Dictyostelium and Human Disease." *American Journal of Human Genetics* 65, no. 1 (1999): 25–30.

Snyder, L. J. *The Philosophical Breakfast Club: Four Remarkable Friends Who Transformed Science and Changed the World*. New York: Broadway Books, 2012.

Storz, J. F. and McClelland, G. B. "Rewiring Metabolism Under Oxygen Deprivation." *Science* 356, no. 6335 (2017): 248–9.

Sugahara, M. et al. "Differences in Heat Sensitivity Between Japanese Honeybees and Hornets Under High Carbon Dioxide and Humidity Conditions Inside Bee Balls." *Zoological Science* 29, no. 1 (2012): 30–6.

Sugahara, M. and Sakamoto, F. "Heat and Carbon Dioxide Generated by Honeybees Jointly Act to Kill Hornets." *Naturwissenschaften* 96, no. 9 (2009): 1133–6.

Tan, K. et al. "Wasp Hawking Induces Endothermic Heat Production in Guard Bees." *Journal of Insect Science* 10 (2010): 142.

Tokuda, Y. "Studies on the Honey Bee, with Special Reference to the Japanese Honey Bee." *Transactions of the Sapporo Natural History Society* 11 (1924): 1–27.

Ugajin, A. et al. "Detection of Neural Activity in the Brains of Japanese Honeybee Workers During the Formation of a 'Hot Defensive Bee Ball.'" *PLoS One* 7, no. 3 (2012): e32902.

Weiss, K. M. "'Nature, Red in Tooth and Claw,' So What?" *Evolutionary Anthropology* 19 (2010): 41–5.

West, S. A. et al. "Evolutionary Explanations for Cooperation." *Current Biology* 17, no. 16 (2007): R661–72.

Wheeler, W. M. "The Ant-Colony As an Organism." *Journal of Morphology* 22, no. 2 (1911): 307–25.

Wilson, E. O. *Sociobiology: The New Synthesis*. Cambridge, MA: Belknap Press, 2000.

Wilson, E. O. *The Insect Societies*. Cambridge, MA: Belknap Press, 1971.

Wilson, E. O. and Gómez Durán, J. M. *Kingdom of Ants: José Celestino Mutis and the Dawn of Natural History in the New World*. Baltimore, MD: Johns Hopkins University Press, 2010.

Wilson, E. O. and Hölldobler, B. "Eusociality: Origin and Consequences." *Proceedings of the National Academy of Sciences* USA 102, no. 38 (2005): 13367–71.

Xavier, J. B. "Sociomicrobiology and Pathogenic Bacteria." *Microbiology Spectrum* 4, no. 3 (2016).

Zimmer, C. A. "Weakness in Bacteria's Fortress." *Scientific American* 312 (2015): 40–5.

Zimorski, V. et al. "Endosymbiotic Theory for Organelle Origins." *Current Opinion in Microbiology* 22 (2014): 38–48.

POSTSCRIPT

Blackiston, D. J. et al. "Retention of Memory Through Metamorphosis: Can a Moth Remember What It Learned as a Caterpillar?" *PLoS One* 3, no. 3 (2008): e1736.

Davidson, D. W. et al. "Histology of Structures Used in Territorial Combat by Borneo's 'Exploding Ants.'" *Acta Zoologica* (Stockholm) 93 (2011): 487–91.

De Loof, A. et al. "The Essence of Insect Metamorphosis and Aging: Electrical Rewiring of Cells Driven by the Principles of Juvenile Hormone-Dependent Ca(2+)-Homeostasis." *General and Comparative Endocrinology* 199 (2014): 70–85.

Downey, P. "Profile of Sérgio Ferreira." *Proceedings of the National Academy of Sciences* USA 105, no. 49 (2008): 19035–7.

Hershkowitz, M. and Samuel, D. "The Retention of Learning During Metamorphosis of the Crested Newt (*Triturus Cristatus*)." *Animal Behaviour* 21, no. 1 (1973): 83–5.

Hölldobler, B. and Wilson, E. O. *The Leafcutter Ants: Civilization by Instinct.* New York: W. W. Norton & Company, 2010.

Ishay, J. S. "Hornet Flight Is Generated by Solar Energy: UV Irradiation Counteracts Anaesthetic Effects." *Journal of Electron Microscopy* (Tokyo) 53, no. 6 (2004): 623–33.

Jackson, D. E. "Nutritional Ecology: A First Vegetarian Spider." *Current Biology* 19, no. 19 (2009): R894–5.

Jackson, D. E. "Social Spiders." *Current Biology* 17, no 16 (2007): R650–2.

Jones, T. H. et al. "The Chemistry of Exploding Ants, *Camponotus Spp.* (Cylindricus Complex)." *Journal of Chemical Ecology* 30, no. 8 (2004): 1479–92.

King, G. F. "Venoms to Drugs: Translating Venom Peptides into Therapeutics." *Australian Biochemist* 44, no. 3 (2013): 13–15; *Expert Opinion on Biological Therapy* 11, no. 11 (2013): 1469–84.

Levin, M. "Molecular Bioelectricity: How Endogenous Voltage Potentials Control Cell Behavior and Instruct Pattern Regulation In Vivo." *Molecular Biology of the Cell* 25, no. 24 (2014): 3835–50.

Pai, V. P. et al. "Genome-wide Analysis Reveals Conserved Transcriptional Responses Downstream of Resting Potential Change in *Xenopus* Embryos, Axolotl Regeneration, and Human Mesenchymal Cell Differentiation." *Regeneration* (Oxford) 3, no. 1 (2015): 3–25.

Plotkin, M. et al. "Solar Energy Harvesting in the Epicuticle of the Oriental Hornet (*Vespa Orientalis*)." *Naturwissenschaften* 97, no. 12 (2010): 1067–76.

Smith, C. G. and Vane, J. R. "The Discovery of Captopril." *FASEB Journal* 17, no. 8 (2003): 788–9.

Tully, T. et al. "Memory Through Metamorphosis in Normal and Mutant *Drosophila.*" *Journal of Neuroscience* 14, no. 1 (1994): 68–74.

Waheed, H. et al. "Snake Venom: From Deadly Toxins to Life-saving Therapeutics." *Current Medicinal Chemistry* 24, no. 17 (2017): 1874–91.

INDEX

A

adenosine triphosphate (ATP), 36–37, 151, 195

Africa, 14, 48, 50, 61–62, 83, 89, 95, 137, 167

Africanized honeybees, 167

Age of Wonder, 48, 53–54, 168–171

Age of Wonder, The (Holmes), 169n

agility, 15

aging, 41–42, 66

air, electricity and, 59

Alexander, Richard D., 165

Alien vs. Predator, 140n

Aliens, 150n

altruism, 146

amber preservation, 11, 143

Ambystoma mexicanum, 197

amino acids, 35, 39

Amoeba proteus, 153–155

amoebas, 153–155, 156

amphibians, 17, 81–84
    axolotls, 197
    metamorphosis and, 197
    neoteny, 197
    newts, 17, 18–22, 83
    salamanders, 64

ampullae of Lorenzini, 62

anaphylactic shock, 110n

ancestors, 9

ancient legends, 50–51, 73, 103–105

anesthesiology, 52–53, 101, 107, 108

Animal Weapons: The Evolution of Battle (Emlen), 4

anteaters, 154

antelopes, 14–15

antibiotic resistance, 8, 157–159

anticoagulants, 94, 107–108

antidotes, 16, 131

ants, 162–164, 173–174, 209

Aparasphenodon brunoi, 83–84

aphids, 163, 164

Apis cerana japonica, 167

aposematic coloration, 81–82

arachnids, 87

Argentina, 53

Aristotle, 50, 169

arms races, 13–20, 41

arthropods, 87–88

Asia, 50, 83, 102

Asian giant hornets, 167

astronauts, 119, 121

atmosphere, 85

atoms, 25–26, 37, 57

ATP (adenosine triphosphate), 36–37

auger snails, 130

Australia, 100, 138

Austroplatypus incompertus, 163

autopoiesis, 37

autotrophs, 36

axolotls, 197

Aysheaia, 143

B

bacteria
    aging process of, 39–40

antibiotic resistance and, 8, 157–159
arms races, 15
botulinum toxin producing, 111
evolution of, 8
fossil record and, 11
Komodo dragons and, 73–74
leeches and, 93–94
life of, 27–28
myxobacteria, 156–157
quorum sensing and, 159
Rickettsia, 152n
social, 156
Staphylococcus aureus, 157–158
X-, 154–155, 156
Bald's Leechbook, 91
Baldscientist, 4
Banda, 137
barracudas, 148
batrachotoxins, 97
bats, 60, 107–108
Beck, Dr. Daniel D., 73n
bee balls, 166–168
bees, 162, 166–168
beetles, 97
Bernabé, Father Cobo, 54
Bernard, Claude, 106–107
Bible, 96
biochemistry
    caddisflies and, 22
    Komodo dragons and, 74
    non-active DNA and, 12
    rays, 47
    toxins and, 76
biodiversity, 9, 11, 146
bioelectricity, 46, 56–58, 63, 66–69, 197
biofilms, 158–159
Biogenesis: Theories of Life's Origin (Lahav), 24n
biological evolution. *See* evolution
birds
    blue-capped ifrit, 96
    European migratory quail, 95
    pitohui, 96–97
    spoor-winged goose, 97–98
    toxic, 94–98
    venomous, 84
Blarina brevicauda, 102
blarina toxins, 102
bloodletting, 91–92
blue-capped ifrit, 96
blue-ringed octopus, 17
Bonner, Dr. John Tyler, 160
Bothrops jararaca, 200–201
botox, 110–112
botulinum toxins, 110–112
Braude, S., 165n
Brazil, 167n
Brazilian pit vipers, 200–201
British Columbia, 141
Brodie, Dr. Edmund III, 19
Brodie, Edmund D., Jr., 18–19
Brodie, Sir Benjamin C., 106
Burgess Shale, 141–143

C

caddisflies, 21–22
calculus, 117n
California Institute of Technology, 131n
California newts, 18–22
Camargo, Antonio, 118n
Cambrian Explosion, 141
Camponotus saundersi, 209
Canarian shrews, 102
cantharidin, 97
Captopril, 201
carbon, 25n
Caron, Jean-Bernard, 142
carrion, 111
Catania, Dr. Kenneth, 55
cavitation, 125–126
cells
    aggregation of, 160
    bioelectrical properties of, 67–68
    cell membranes, 56

chemical equilibrium of, 56–57
cnidocysts, 120
complexity of, 28
contamination of, 153–155
death of, 41
definition of, 27–29
differentiation, 41
dissociation of, 64, 65
DNA and, 34–35
electricity and, 56–58
endosymbiosis, 151–153, 155, 156
excitable, 58
nematocysts, 120
polarization of, 57
symbiosis of, 151
viruses and, 26–27
voltage of, 57–58
Central America, 72, 81, 107–108
cheetahs, 14–15
chemical equilibrium. See death
Chicxulub crater, 118n
children, 133n
chimpanzees, 72
Chondrodendron tomentosum, 106–107
chromosomes, 32
Chrondomyces crocatus, 156–157
Clark, Craig, 132
cleaner fish, 148
cnidarians, 64, 119–122. See also
corals; hydrae; jellyfish; sea
anemones
cnyidocysts, 120
Cobo, Father Bernabé, 54n
cobras, 131
coelacanths, 62–63
coevolution, 14–15, 19
Cold War, 14
color perception, 126–128
coloration, aposematic, 81–82
Columbia University, 49
commensalism, 147
common cold, 27
common medicinal leeches, 92

competition, 145–150
cone snails, 80–81, 128–139
conotoxins, 130–132, 136n, 138
consciousness, 5
Conus geographus, 131, 137–138
Conus gloriamaris, 136
Conus purpurascens, 136
conus snails, 130
Conus striatus, 133–134, 138
Conus textile, 136
cooperative behaviors, 27, 150–155, 156–157, 170
corals, 64, 83, 103–106, 119–122
Cornell University, 136n
Corythomantis greeningi, 83–84
coturnism, 95
Crick, Sir Francis, 169–170
Crocidura canariensis, 102
crocodile newts, 83
crocodiles, 147–148
crustaceans
mantis shrimp, 123–128
remipedes, 88
venomous, 87–88
water fleas, 33, 144
Cruz, Dr. Lourdes, 133
CSI: Miami, 136
ctenophores, 122–123
Cuba, 103
curare plant, 106–107, 111

D

Daphnia, 33
Darwin, Charles, 1–2, 5, 12, 39, 54, 61, 145n, 162n, 171
Davis, Dr. Wade, 17–18
Dawkins, Richard, 9
death
awareness of, 7
mysteriousness of, 40–42
poisons and, 76
venoms and, 16, 18, 132, 137–138

decay, 11
decomposers, 11
defensive strategies
    electricity, 58, 60
    limited nature of, 99
    social behavior, 167
    startle response, 130
    toxins, 22, 77–83, 88, 129
    venoms, 88, 98, 100–101, 110
Democritus, 24, 25
dental work, 93
depression, 132–133
Desmodus rotundus, 107–108
Diaaemus youngi, 107–108
diabetes, 73, 81
Dictyostelium, 160
differentiation, 41
dinosaurs
    conotoxins and, 136–137
    extinction of, 10–11, 20, 117–118
    Sinornithosaurus, 74–75
    Uatchitodon, 75
    venomous, 71–75
Diphylla ecaudata, 107–108
DNA, 131n
    cells and, 34–35
    non-active, 12
    sequencing of, 32
    variation and, 29–32
Dobzhansky, Theodosius, 36
Doctor Who, 66n
dolphins, 60, 173
Draculin, 108
d-tubocurarine, 106
Dworkin, Dr. Martin, 156–157

E

Earth
    life on, 25, 84–87
    origin of life and, 38–40
Ebola, 27
echidnas, 99–101
Echinotriton andersoni, 83

Echinotriton chinhaiensis, 83
echolocation, 60
Einstein, Albert, 44, 115
electric eels, 54n
electric fish
    ancient medicinal uses of, 51–53
    catfish, 47–50
    eels, 53–56, 54n
    electrogenic field of, 59–60
    elephant fish, 61
    rays, 50–53
electric organs, 55–56, 61–62
electrical fields, 59
electricity, 43–46. *See also* electric fish
    air and, 59
    detecting, 58–63
    genes and, 63
    life and, 67–69
    living organisms and, 58–59
    water and, 59
electromagnetism, 44, 126–127
electrons, 45, 46, 57
Electrophorus electricus, 53, 55
electroreception, 59–62
elephant fish, 61
Emlen, Dr. Douglas J., 4
endangered animals, 92, 166
Endean, Dr. Robert, 138–139
endosymbiosis, 151–153, 155, 156
environment
    acquiring energy from the, 2
    bioelectricity and, 69
    changes in the, 19–22
    gene expression and, 63
    mutations and the, 29
    toxins acquired from the, 82, 95, 97–98
    venoms and, 78–79
enzymes, 35, 79, 157
epigenetics, 13
Erpeton tentaculatum, 204–205
Euglena, 154
eukaryotes, 32, 151–155
Europe, 83, 95, 102

European honeybees, 167
European migratory quail, 95
eusociality, 163–166
evolution
    antibiotic resistance and, 8
    arms races, 13–20, 41
    bacteria and, 8
    coevolution, 14–15, 19
    endosymbiosis and, 155
    epigenetics and, 13
    genetic drift and, 13
    horizontal gene transfer and, 13
    human, 9
    life and, 25, 115
    microbes and, 15–20
    molecules and, 13, 15–20
    mutations and, 12–13
    offensive and defensive strategies
        and, 7
    organic, 7
    real-world applications, 8
    reproduction and, 7, 26
    survival strategies, 8–9
    symbiosis and, 13
    toxins and, 15
extinction, 9–10, 20
extraterrestrial life, 38–40

F

fire ants, 154
fish
    cleaner, 148
    coelacanths, 62–63
    electric, 47–50, 51–56, 59–61
    elephant, 61
    pilot, 148
    puffer, 16–17, 173
    zebrafish, 64
flatworms, 12, 17, 65–66, 123, 164
Florida Institute of Technology, 88
fluke flatworms, 164
flying death, 106
folklore. See ancient legends
food, avoid becoming, 2, 58

Food and Drug Administration, 93
Formica yessensis, 173–174
fossil record, 10–11
France, 92–93
Franklin, Benjamin, 51
frogs
    poison arrow, 17, 81–84
    venomous, 83–84
fruiting bodies, 156, 160–161
Fry, Dr. Bryan, 77
fungi, fossil record and, 11

G

Galen, 50
Galvani, Luigi, 52, 56
garter snakes, 18–22
gazelles, 14–15
genes, 32, 63, 68
genetic code, 35
genetic drift, 13
genetic information
    passing on, 2, 20
    venoms and, 78–79
genetics, 20
genomes, 33, 69
genotype, 33–34
Gila monsters, 71–73, 81
Gilbert, William, 44
Gondwana, 143
gout, 52
great white sharks, 148
Greenberg, Everett, 157
grizzly bears, 193–194
grolar bears, 193–194
Guatemala, 72

H

Haldane, J. B. S., 26, 39
Hallucigenia, 139–141
Harris-Warrick, Professor Ron,
    136n
Harvard University, 157
Hawaii, 103

Hawaii Five-O, 136
heavy metals, 76
Hebrew University of Jerusalem,
    24n
Heloderma horridum, 71–72
Heloderma suspectum, 71–72
hemlock seeds, 95–96
Heraclitus, 113n
heredity, 32
hermit crabs, 147
Hermitte, Dr. Louis C. D., 137–138
Heterocephalus glaber, 165
heterodonts, 74–75
heterotrophs, 36
Himasthla, 164
Hippocrates, 92
Hirudo medicinalis, 92
Hispaniola, 103
hoarding, 102
Hokkaido Island, 173
Holmes, R., 169n
homobatrachotoxin, 97
honeybees, 166–168
hooded pitohui bird, 96–97
Hopkins, Chris, 133
horizontal gene transfer, 13
hornets, 167–168, 173
humans, 15, 149
hydrae, 64–66, 67, 119–122
hypodermic needles, 79–80, 130

I

Iberian newts. *See* ribbed newts
identical twins, 29–30
Ifrita kowaldi, 96
immortality, 63–67
immunity, 111, 123
impurities, 59
"In Memoriam A.H.H."
    (Tennyson), 145
indigenous cultures, 73, 103–105
Indonesia, 62, 73
infection
    bacteria and, 94, 152, 154–155

MRSA, 157–158
    non-active DNA and, 12
information transfer, life and, 25
infrared, 127
Innocent II, Pope, 124n
insecticides, 78
insects, 87
    Africanized honeybees, 167
    ants, 162–164, 173–174, 209
    aphids, 163, 164
    Asian giant hornets, 167
    bees, 162
    beetles, 87
    European honeybees, 167
    fossil record and, 11
    honeybees, 166–168
    hornets, 167–168, 173
    Japanese honeybees, 167–168
    metamorphosis and, 197
    wasps, 168
    weevils, 163
insulin, 135
intelligence, coevolution and, 15
*International Journal of
    Developmental Biology, The*, 69n
ions, 57
Ito, F., 107n

J

Jahren, Hope, 3n
Japan, 173
Japanese honeybees, 167–168
jellyfish, 64, 66–67, 119–123
Jenner, Dr. Ronald, 88
Jeon, Dr. Kwang, 153–155
Jurassic Park, 136

K

killer snails, 128–139
King, Dr. Harold, 106
knife fish. *See* electric fish, eels
Kohn, Dr. Alan J., 138
Komodo dragons, 72–74

L

Lab Girl (Jahren), 3n
laboratory animals, 18–19
Lahav, Noam, 24n
Largus, Scribonius, 52
Latimeria chalumnae, 62–63
Latimeria menadoensis, 62
leafcutter ants, 164
leech therapy, 91–93

M

Madani, G, 110n
magnetism, 44
Mahé Island, 137
Malapterurus electricus, 47–50
Malpighi, Marcello, 61–62
mammals. *See also* specific types of
    mammals
    bats, 107–108
    echidnas, 99–101
    mole rats, 164–166
    naked mole rats, 165
    platapuses, 99–101
    shrews, 102–103
    slow lorises, 108–110
    solenodons, 102–103
    venomous, 84, 98–110
Mandriota, Dr. Frank, 49–50
mantis shrimp, 123–128
Margulis, Dr. Lynn, 151–152
matter, 5
Maxwell, James Clerk, 44
medications
    antibiotic resistance, 8, 157–159
    blood pressure, 200–201
    Captopril, 201
    children and, 133
    toxins and, 80–81
    venoms and, 73, 80–81, 91,
        200–201
Megalania, 74
Melyrid beetle, 97
Mendeleev, Dmitri, 39

Mesopotamia, 136
metabolism, 37–38
metamorphosis, 197
Mexican beaded lizards, 71–72
Mexico, 72, 118
microbes, 15–20, 153–155
microorganisms, 35, 40, 86
Miller, Stanley, 39
mole rats, 164–166
molecules, 13, 15–20, 26
monotremes, 99–101
moray eels, 46n, 148
Mormyridae, 49, 61
Morocco, 83
Morris, Simon Conway, 142
Mother Nature Is Trying to Kill You
    (Riskin), 4
movement, 113–114, 115–116
MRSA, 157–158
multicellular organisms, 28
    death and, 41
    development of, 152–153
    nervous systems of, 161
    oxygen and, 85
    slime molds, 159–161
mutations, 12–13, 20, 29–34
mutualism, 147–148
myriapods, 87
myxobacteria, 156–158
Myxococcus xanthus, 156

N

naked mole rats, 165
Narcine baliensis, 50–53, 52
nárkē, 52–53
natural disasters, 13
Natural History Museum, London,
    88
natural selection, 12, 20
Naucrates ductor, 148
Navajo tribe, 73
Nekaris, K. A., 98n, 110n
nematocysts, 120
Neomys anomalous, 102

Neomys fodiens, 102
nerve transmission, 116
Nesophontes edithae, 103
neurobiological effects, 132, 136
neuropharmacology, 107
neutrons, 46n
New Guinea, 96
New York Times, 49–50
Newton, Sir Isaac, 117n, 122
newts
    California, 18–22
    crocodile, 83
    ribbed, 83
    toxins and, 17, 18–22
nightmares, 139
nonliving matter, 26
North America, 102
Northern Arizona University, 165
nudibranchs, 122–123
numbness, 48, 51, 52, 93, 108, 137
Nycticebus, 108–110

O

offensive strategies, 4, 7, 58
Olivera, Dr. Baldomero M., 130–134
On the Magnet (Gilbert), 44
On the Origin of Species (Darwin), 61
onychophorans, 143
Oparin, Aleksandr, 39
Origin of the Species (Darwin), 145
Ornitorynchus anatinus, 99–101
ourari, 106
Oxford University, 45
oxidization, 85–86
oxygen, 84–87
Oxygen Holocaust, 86
ozone layer, 86–87

P

Pagán, Oné R., 69n
pain therapy, electric fish and, 52
Palythoa toxica, 104–106

palytoxins, 105–106, 111
pangolins, 99
Parabuthus transvaalicus, 89–90
Paracelsus, 76
paralysis, 106, 132, 138
parasites, 15, 77, 90–94, 97, 110, 147,
    148, 164
Paris japonica, 33
Parsek, Dr. Matthew, 157
Paxton, Bill, 150n
peacocks, 81
Pemphigus, 163
Penfield, Glen, 118n
Peripatoides indigo, 143–144
peripatus, 143–144
Peru, 54n
pharmacology, 107
phenotypes, 34
pheromones, 101n
Philippines, 130–131
Philosophical Breakfast Club, The
    (Snyder), 169n
photons, 126–127
photoreceptors, 127
photosynthesis, 86, 151n, 195
Physarum, 160
physical therapy, 52
physics, laws of, 25
physiology, 12, 35n, 107
pilot fish, 148
pit viper snakes, 84, 200–201
pitohui birds, 96–97
Pitohui dichrous, 96
pizzly bears, 193–194
"Planaria: an animal model that
    integrates development,
    regeneration and
pharmacology" (Pagán), 69n
planarian worms, 12, 65–69, 144,
    145n
plants
    behavior of, 29
    carbon dioxide and, 195
    curare, 106–107, 111

DNA and, 33
  photosynthesis, 36
  survival of, 3
  toxins and, 77
Plato, 50
platypuses, 59, 99–101
Pleurodeles waltl, 83
plovers, 147–148
poison arrow frogs, 17, 81–84
poisons, definition of, 76
polar bears, 193–194
polarized light, 128
polyclads, 123
Pompeii, 51
Portugal, 83
predators. *See also* specific predators
  arms races, 14–16, 20
  electroreception and, 59–60
  immobilizing prey, 53
  inducing fatigue in prey, 53
  numbing prey, 48, 51, 52, 93,
    108, 137
  "sit-and-wait," 204–205
  smell and, 144
  speed and, 122, 125
  toxins and, 77–78
  tracking prey, 53
prey. *See also* specific prey
  arms races, 14–16
  exploding, 209
  heightened senses of, 59
  paralysis of, 106, 132, 138
  speed and, 122, 125
"primordial soup" theory, 39
Princeton University, 160
Pringle, Sir John, 51
proboscis, 134–135, 138
prokaryotes, 32, 151–152, 156
proteins
  amino acids and, 35, 39
  DNA and, 34–35
  hirudin, 94
protons, 46, 57
puberty, 31

Puerto Rico, 103
puffer fish, 16–17, 173

Q

quail, 95
quorum sensing, 158–159, 160

R

Raleigh, Sir Walter, 106
Ramskold, Lars, 142
raptorial appendages, 124
rattlesnakes, 84
rays, 62
reflexes, 116
regeneration, 63–66
remipedes, 88
reproduction
  behavior and, 29–30
  eusociality and, 163, 165
  evolution and, 7
  life and, 26
ribbed newts, 83
Rickettsia, 152n
Riskin, Dr. Daniel, 4
River Out of Eden (Dawkins), 9
rocket launches, 118–119
Rocky Mountain spotted fever, 152n
Rode-Margono, J. E., 98n
Rumphius, Georg Eberhard, 137

S

sacrifice, 146, 209
saguaro wine festival, 73
salamanders, 64
scavengers, 11
Schrödinger, Erwin, 24
scorpions, 89–90
Scytodes, 139–141
sea anemones, 64, 119–122, 147
sea slugs, 122–123
sediment, 11

self-maintenance. *See* autopoiesis
senescence, 41–42, 66
Serpent and the Rainbow, The (Davis),
    17–18
Seychelles islands, 137
shamans, 73
sharks, 62, 148
shrews, 102–103
shrimp, 88, 148, 163
Sinornithosaurus, 74–75
skates, 62
Skeptical Enquirer, 18
slime molds, 159–161
slow lorises, 108–110
Smith, Martin R., 142
snails, 90
    auger, 130
    cone, 80–81, 128–139
    conus, 130
    turrids, 130
snakes
    lizards relation to, 72
    pit vipers, 84, 200
    rattlesnakes, 84
    tentacled, 204–205
    venomous, 71
Snyder, L.J., 169n
sociomicrobiology, 157–161
Solenodon cubanus, 103
Solenodon paradoxus, 103
solenodons, 102–103
South America, 53, 54, 81–82, 106,
    107–108
Spain, 83
speed
    cnidarians, 119–122
    coevolution and, 14–15
    hallucigenia, 141–144
    killer snails, 128–139
    mantis shrimp, 123–126
    spiders and, 139–141
    spitting spiders, 139–141
    tentacled snakes and, 204–205
    time and, 115–119
Speleonectes tulumensis, 88

Spencer, Herbert, 145n
Spider-Man, 139n, 144n
spiders, 139–141, 173
spiritual beliefs, 44
spitting spiders, 139–141
sponges, 64–65
spoor-winged goose, 97–98
Stanford University, 131n
Star Trek: Nemesis, 31n
Star Trek Beyond, 120
startle reflex, 130, 135
Steno, Nicolas, 61–62
Stensen, Niels. *See* Steno, Nicolas
Stockholm Syndrome, 150–155
stomatopods, 123–126
substance abuse, 78
supernatural beliefs, 44
superorganisms, 162–163
survival. *See also* coevolution
    instinct, 2, 3, 7, 8–9, 14–15
    mechanisms for, 34, 145–150,
        146, 150–155
survival of the fittest, 145
symbiosis, 13, 147, 151
Synalpheus regalis, 163

T

tardigrades, 142
Taricha granulosa, 18–22
taxonomy, 12
technology, 149
Tennyson, Alfred Lord, 145
tentacled snakes, 204–205
termites, 164
tetrachromatic vision, 128
tetrodotoxin (TTX), 16–22, 132,
    173
Tetronarce cowleyi, 50–53, 52
Thamnophis sirtalis, 18–22
Thaxter, Roland, 156
Thor, 44
thought, 116
thunderstorms, 43–44, 58
time, 5, 115–116

Tohono O'odham Native
    Americans, 73
toxins, 15
    batrachotoxins, 97
    birds, 94–98
    blarina, 102
    botulinum, 110–112
    conotoxins, 130–132, 136, 138
    definition of, 76–78
    homobatrachotoxin, 97
    immunity to, 111
    natural context of, 110–112
    origin of the word, 76n
    oxygen, 87
    palyotoxins, 105–106, 111
    poison arrow frogs, 17, 81–84
    tetrodotoxin (TTX), 16–22, 132,
        173
    venomous crustaceans, 87–88
traits, 34
Transvaal thick- or fat-tailed
    scorpions, 89–90
Tuatara lizard, 74
turrid snails, 130
Turritopsis nutricula, 66–67
typhus, 152n

U

Uatchitodon, 75
ultraviolet light, 126–127
ultraviolet radiation, 87
unicellular organisms, 27–28,
    153–155
uniqueness, 31–32
United States, 72, 167
University of California at Davis,
    89–90
University of Chicago, 39
University of Michigan, 165
University of Minnesota, 156
University of Philippines, 131n
University of Queensland, 138
University of Tennessee, 153
University of Utah, 130, 131

University of Washington, 138
Urey, Harold, 39

V

Vanderbilt University, 55
varanids, 72, 73–74
variation, 31–32
Vaughan, Terry, 165
velvet worms, 141–144
venoms, 16
    bats, 107–108
    cnidarians, 119–122
    crustaceans, 87–88
    definition of, 78–81
    dinosaurs, 71–75
    immunity to, 123
    leeches, 90–94
    mammals, 84, 98–110
    medications and, 80–81, 91,
        200–201
    natural context of, 110–112
    pre-, 89–90
    scorpions, 89–90
    snails, 128–139
    spiders, 139–141
Vermeer, Johannes, 136
Vespa mandarinia, 167
viruses, arms races, 15
visual acuity, coevolution and, 15
Volta, Alessandro, 52, 56
voltage, 57–58
von Humboldt, Alexander, 48,
    54–55
von Reumont, Dr. Björn M., 88
vultures, 111

W

Walcott, Dr. Charles Doolittle,
    141–142
"warm little pond" theory, 39
wasps, 168
water
    electricity and, 59

poisonous effects of, 76
water bears, 143
water fleas, 33, 144
weevils, 163
What is Life? (Schrödinger), 24
What Mad Pursuit (Crick), 169–170
Wilson, Dr. Edward O., 82, 157
worms
flatworms, 12, 17, 65–66, 123
   fluke flatworms, 164
   Hallucigenia, 141–144
   Himasthla, 164
   leeches, 90–94
   planarian, 12, 65–69, 144, 145n
   polyclads, 123
   velvet, 141–144
Wren, Christopher, 79–80

X

X-bacteria, 154–155, 156
Xianguang, Hou, 142
Xibalbanus tulumensis, 88

Y

Yager, Dr. Jill, 88
Yucatán Peninsula, 118n

Z

zebrafish, 64
Zeus, 44
zombies, 17
zooanathids, 104

ABOUT THE AUTHOR

DR. ONÉ R. PAGÁN is a husband and a father, as well as a biology professor, scientist, blogger, and writer. He is in absolute awe of the natural world, especially the fact that it can be understood through science and mathematics. He spends quite a bit of his time explaining science to family, students, and essentially anyone within earshot. He also loves learning about nature firsthand through his scientific research. He has published original work in various scientific journals including the International Journal of Developmental Biology, Neuroscience Letters, Toxicon, Neurochemical Research, and Pharmacology, Biochemistry, and Behavior among others. He holds an undergraduate degree in Natural Sciences and a Master's degree in Biochemistry, both from the University of Puerto Rico, and a Doctorate in Pharmacology with a strong emphasis in neurobiology from Cornell University.